Corrosion - Fundamentals and Protection Mechanisms

Edited by Fahmina Zafar, Anujit Ghosal and Eram Sharmin

Published in London, United Kingdom

IntechOpen

Supporting open minds since 2005

Corrosion - Fundamentals and Protection Mechanisms
http://dx.doi.org/10.5772/intechopen.94634
Edited by Fahmina Zafar, Anujit Ghosal and Eram Sharmin

Contributors
Duduzile Zamavezi Nkomo, Nomsombuluko Masia, Burak Tüzün, Şaban Erdoğan, Ntebogeng Mogale, Walace Matizamhuka, Prince Cobbinah, Dwarika Prasad, Mykhaylo Viktorovych Yarmolenko, Lukman O. Olasunkanmi, Fahmina Zafar, Shumaila Masood, Anujit Ghosal, Eram Sharmin, Nahid Nishat

First published in London, United Kingdom, 2022 by IntechOpen
IntechOpen is the global imprint of INTECHOPEN LIMITED, registered in England and Wales, registration number: 11086078, 5 Princes Gate Court, London, SW7 2QJ, United Kingdom
Printed in Croatia

British Library Cataloguing-in-Publication Data
A catalogue record for this book is available from the British Library

Additional hard and PDF copies can be obtained from orders@intechopen.com

Corrosion - Fundamentals and Protection Mechanisms
Edited by Fahmina Zafar, Anujit Ghosal and Eram Sharmin
p. cm.
Print ISBN 978-1-83968-605-4
Online ISBN 978-1-83968-606-1
eBook (PDF) ISBN 978-1-83968-607-8

For EU product safety concerns:
IN TECH d.o.o., Prolaz Marije Krucifikse Kozulić 3, 51000 Rijeka, Croatia, info@intechopen.com or visit our website at intechopen.com.

Meet the editors

Dr. Fahmina Zafar is a senior researcher in the Department of Chemistry, Jamia Millia Islamia (JMI), New Delhi, India. She received her MSc and Ph.D. in Chemistry from JMI in 1999 and 2006, respectively. She has worked as a principal investigator for the Department of Science and Technology, Women Scientists Scheme, Government of India; a postdoctoral fellow for University Grant Commission India; and a research associate in the Scientist's Pool Scheme of the Council of Scientific and Industrial Research, India. Dr. Zafar has four patents, fifty-six peer-reviewed papers in reputed journals, eight books, and twenty-seven book chapters with international publishers to her credit. She has presented more than fifty research papers at national and international conferences. Her research work involves the development of bio-based polymers, metallopolymers, organic-inorganic hybrids, coordination polymers, and nanocomposites for the green environment in different fields including adsorption, antimicrobial, and corrosion-protective applications.

Dr. Anujit Ghosal is a researcher at Richardson Centre for Functional Foods and Nutraceuticals, University of Manitoba, Canada. He obtained a BSc with honors in Chemistry from the University of Delhi, India in 2007. He obtained his MSc in Materials Chemistry and Ph.D. in Chemistry from Jamia Millia Islamia (JMI), India in 2009 and 2015, respectively. He has worked as a national postdoctoral fellow under SERB fellowship in the School of Biotechnology, Jawaharlal Nehru University, India, and served as an assistant professor for one year at Galgotias University, India. He has published in various internationally reputed refereed journals and participated in many national and international workshops. His research expertise involves polymer chemistry, valorization of agri-products, hydrogels, nanomaterials, drug delivery, and electrochemical corrosion.

Dr. Eram Sharmin is an associate professor in the Department of Pharmaceutical Chemistry, College of Pharmacy, Umm Al-Qura University, Makkah, Saudi Arabia. She obtained a BSc in Chemistry and MSc in Organic Chemistry from Aligarh Muslim University (AMU), Aligarh, India in 1998 and 2000, respectively. She obtained her Ph.D. in Chemistry from Jamia Millia Islamia (JMI), India in 2007. She was previously a senior research associate at the Council of Scientific and Industrial Research (CSIR), New Delhi, India, and a senior research fellow in the Department of Chemistry, Materials Research Laboratory, JMI. She has published books and more than fifty publications in peer-reviewed journals. She has presented more than thirty research papers at national and international conferences. Her research interests include the development of "green" materials such as antimicrobial and protective coatings, films, hydrogels, and packaging materials.

Contents

Preface

Although a natural process, corrosion in materials is responsible for huge direct as well as indirect losses around the world. This book explores the impact of corrosion products on the performance of devices, especially those incorporating metals or metal derivatives. It also discusses corrosion prevention mechanisms, the role of inhibitors in various oil/gas pipeline industries to avoid fatal accidents, and the growth of natural inhibitors over inorganic or organic inhibitors to support the concept of the “green” approach. Furthermore, the book presents the basics of corrosion from thermodynamic and kinetic points of view, discusses the major driving force behind corrosion, and provides insight into possible remediation techniques.

The book is divided into three sections. The first section discusses the fundamentals of corrosion and the driving force behind the process, types, and possible mechanisms involved. The second section provides insight into corrosion inhibitors and the third section discusses the characterization aspects in corrosion sciences with developmental trends associated with computational analysis of corrosion. The impact of computational methodologies like density functional theory and molecular dynamics simulations for the identification of suitable inhibitors is presented. The practical relevance of additives in the manufacturing of alloys using laser powder bed fusion technology is also summarized along with the concept of high-temperature corrosion (hot corrosion). Overall, this book is beneficial for students and researchers who are interested in the field of corrosion sciences, polymer technology, and the designing of inhibitors.

The completion of the book was possible through the collaborative efforts, quick responses, and patience of the staff at IntechOpen, especially Author Service Managers Ms. Nera Butigan and Mr. Darko Hrvojic. We are also grateful to the contributing authors for their excellent chapters. Dr. Fahmina Zafar is thankful to the Department of Chemistry, Jamia Millia Islamia, New Delhi, India for the “Assistant Professor” position and space to work on this publication.

Dr. Fahmina Zafar
Assistant Professor,
Department of Chemistry,
Jamia Millia Islamia,
New Delhi, India

Dr. Anujit Ghosal
Postdoctoral Fellow,
Department of Food and Human Nutrition,
University of Manitoba,
Winnipeg, Canada

Dr. Eram Sharmin
Associate Professor,
Department of Pharmaceutical Chemistry,
College of Pharmacy,
Umm Al-Qura University,
Makkah Al-Mukarramah, Saudi Arabia

Section 1

Introduction and Basics of Corrosion

Chapter 1

Introductory Chapter: Corrosion

Shumaila Masood, Anujit Ghosal, Eram Sharmin, Fahmina Zafar and Nahid Nishat

1. Introduction

Corrosion can also be termed "metallic cancer" because it is an irreversible process that leads to the loss of billions of dollars. It involves the deterioration and consequently loss of solid metallic structure due to the chemical (dry gases, moisture, liquid, ionic solutions, microbes, etc.) or electrochemical (micro-cell formation like Daniel cell) reactions resulting from the potential difference in the structure and presence of a suitable electrolyte (salt-water) [1]. The process of corrosion is a surface phenomenon whose rate depends upon various factors such as temperature, the presence of harmful chemicals in the environment, ionic species, as well as humidity. Once the surface is breached the process of deterioration continues within, thereby making the structure vulnerable to stress and load. The combination with acid in the environment generated via., possible chemical reactions speed up the process resulting in the reduction of the life span of metal structure [2]. Corrosion can be sub-divided into many forms such as pitting, galvanic, intergranular, stress, and others. All these can affect the structural integrity by weakening the metallic units within any construction by corroding rods/wires, water pipelines (leakages), metallic bases, electrical units, etc. Few of these examples are shown in **Figure 1**.

These corrosive reactions change the microstructure of metal which results in the loss of its elasticity, mechanical and tensile strength converting them into flaky and brittle units. The losses incurred due to corrosion directly effects the gross domestic production (GDP) of all the countries. In 2018, the corrosion protection industry was estimated to be around 2.5 trillion (USD) and is expected to cross 3 trillion (USD) by the end of this year [3]. Proper use of corrosion prevention measures can help to refrain from such severe damages and can partially avoid the loss in economy, environmental pollution, direct and other indirect losses associated with corrosion [4].

The most popular remedies involved in corrosion protection are the use of cathodic protection, anodic protection, corrosion inhibitors, and protective coatings. Among them, utilization of corrosion inhibitors (chemical barrier) and protective coatings (physical barrier) is mostly focused on by a larger population around the world [5]. Coatings provide a physical barrier between the harmful environment and the metal surface which enhances the lifespan of metallic structures. A number of coatings that have been developed for corrosion protection include metallic coatings, chemical conversion coatings, organic and inorganic coatings, nanocomposite coatings [6]. However, such coatings have limited applications in complex design and high-rise infrastructures. Thereby coating with rapid curing ability, strong adhesive abilities, and stability against the harsh environment (exposure to atmospheric moisture, UV radiations, and other pollutants) are being researched. On the other hand, corrosion inhibitors are chemical additives that are added in low concentration to the

Figure 1.
Effect of corrosion on different metallic structures.

destructive media, and timely they tend to suppress the corrosion progress [7]. These chemicals are classified as cathodic, anodic, or mixed inhibitors depending on the inhibition of the type of corrosion. The use of renewable resources in the field of corrosion prevention in the form of coatings and inhibitors is gaining significant importance. The chapter describes the types of corrosion, mechanism, and its control through synthetic/renewable resources based on synthetic coatings, micro to nano-coating, and corrosion inhibitors.

2. Corrosion

Corrosion of steel is a spontaneous electrochemical process that takes place in the presence of a solution containing dissolved oxygen. This process includes the delocalization of metal ions through oxidation into the solution at the anode (active area) and mobilization of electrons through reduction of metal to an acceptor such as oxidizing agents, or oxygen or hydrogen ions at cathode (less active area) [8].

The mechanism of corrosion involves the release of electrons from the metallic surface into the electrolyte in the presence of oxygen (**Figure 2**). This process occurs due to the tendency of metals to return to their natural oxidation state. The reactions occurring at both the electrodes can be expressed as follows:

At anode:

$$Fe \longrightarrow Fe^{+2} + 2e^{-} \quad (1)$$

$$Fe^{+2} \longrightarrow Fe^{+3} + e^{-} \quad (2)$$

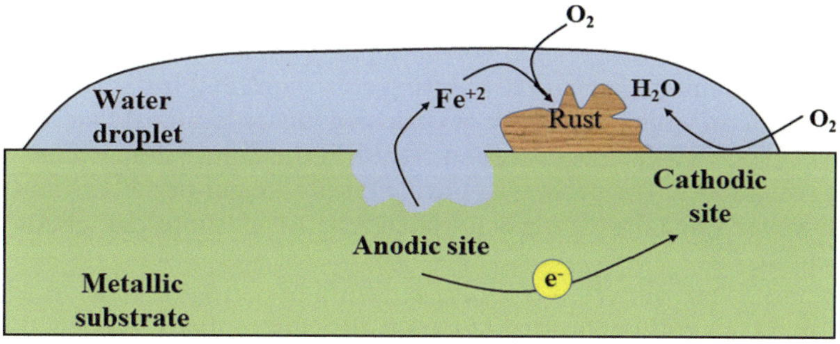

Figure 2.
Mechanism of corrosion.

At cathode:

$$\frac{1}{2}O_2 + H_2O + 2e^- \longrightarrow 2OH^- \tag{3}$$

The deterioration of metallic surfaces depend upon the different forms of corrosion, which is caused due to different type of corrosive environment (**Figure 3**). Dry corrosion, also known as chemical corrosion, is the type of corrosion that occurs in the absence of moisture or water and metal oxidizes only due to the atmosphere. Therefore, this may also be referred to as, an oxidation process sustained by atmospheric oxygen without a liquid solution. At encompassing temperatures, most metals have slow oxidation rates [9]. Galvanic corrosion (GC), also known as bimetallic corrosion, is an electrochemical process where two different

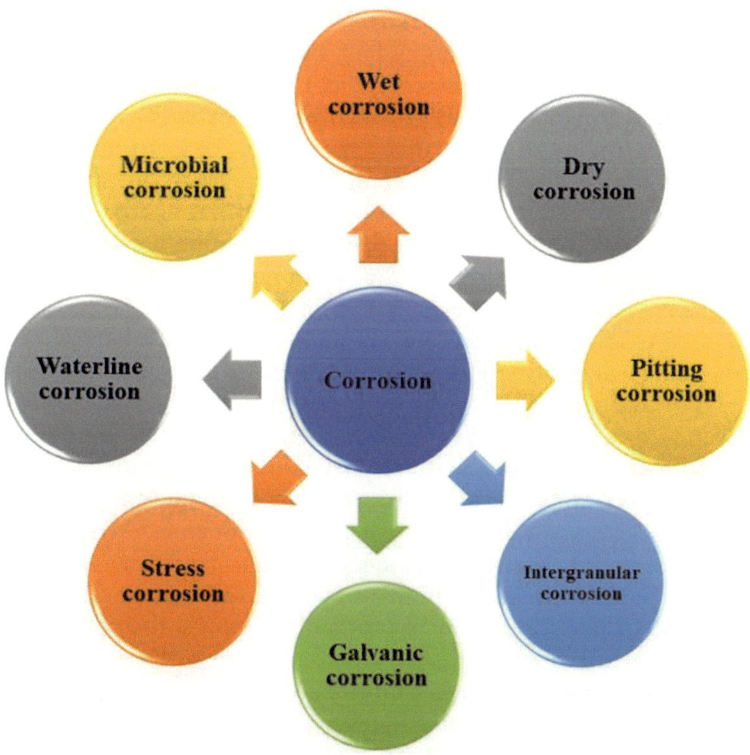

Figure 3.
Different types of corrosion.

metallic materials are connected electrically in a corrosive environment. In this type of corrosion, one metal (the anode) corrodes preferentially while the other metal (the cathode) remains protected [10]. Pitting corrosion (PC) is the most destructive type of corrosion in which the attack of corrosive ions is localized that results in the formation of pits. This type of corrosion results in the failure of machines without much of a weight loss. The process of pitting takes a longer period of time to initiate, however, once the pitting is initiated it penetrates into the section at an accelerated rate [11].

Intergranular corrosion (IC) is a localized form of corrosive attack that is preferentially along the grain boundaries or areas adjacent to them. The corrosion activity occurs at the grain boundary area since it is electrochemically different from the bulk. This process is mainly observed in stainless steel [12]. Waterline corrosion (WC) is an oxidation process that occurs when one part of the metal is submerged in water and another part is in contact with air. Water tanks are often prone to this type of corrosion [13]. Stress corrosion (SC) occurs when tensile stress and corrosive environment work together, often at elevated temperatures. In this type of corrosion stressed area of metal is anodic in respect to the unstressed area of the metal. This corrosion is not visible prior to fracture; therefore, it results in catastrophic failure [14]. Microbiologically induced corrosion (MIC) is a type of corrosion in which metal deteriorates through the metabolic activity of microorganisms. The common bacteria that cause MIC are acid-producing bacteria, sulfate-producing bacteria, and iron-reducing bacteria [15].

2.1 Corrosion protective coatings: combating mechanism

Corrosion cannot be completely eradicated but overcame by using protective coatings and corrosion inhibitors. Protective coatings are developed to retard the corrosion rate to protect metallic substrates [16]. These coatings are applied using several techniques such as roller, moving belt, or brush technique. They are functionalized with the help of various organic, inorganic, hybrid, or metallic layers to enhance their performance [17]. Highly protective coatings provide an impenetrable barrier protecting the substrate from the aggressive environment (**Figure 4**).

Modified polymeric coatings provide enhanced corrosion protection as compared to the simple polymeric coatings [18]. Various types of coating utilized for protection purposes are shown in **Figure 5**. Metallic coating means protecting the metals with the help of metal-coating. These coatings are applied to the substrates for several reasons but among all corrosion protection is major [19]. Various methods are used to apply these coatings on a substrate such as metalizing, electroplating, vapor deposition, hot dipping, cladding, etc. [20]. Depending upon the metal used to coat the substrate, metallic coatings can be divided into two categories: Anodic and cathodic coatings. In anodic coatings, anodes are made up of the alloys that are electrochemically more active than the base metal as a result of which the anodic metal depletes at a faster rate. These anodic coatings act as a physical obstruction between the corrosive environment and the base metal thereby protecting the base metal. These anodic metals can also be called "*Sacrificial anodes*". Zinc, magnesium, and cadmium are well non-sacrificial anodic metals that provide protection to steel [21]. While in cathodic coatings, the coating metal is selected in such a manner that it remains electropositive with respect to the base metal. For example, copper is used to coat steel.

Chemical Conversion coatings, also known as surface passivation, are produced through the chemical and electrochemical reaction of metal. Through these coatings, the surface of a metal is modified such that it possesses desired porosity. These types of coatings are more adhesive as there is a chemical bond and intermediate

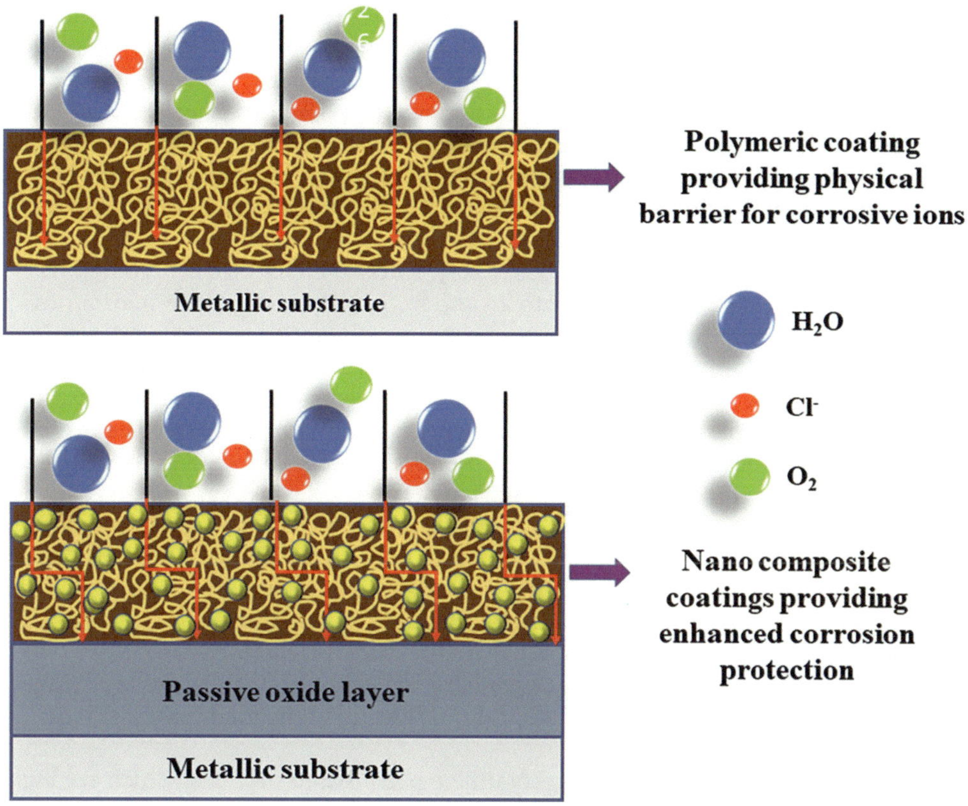

Figure 4.
Corrosion protection mechanism of polymeric and modified polymeric coatings.

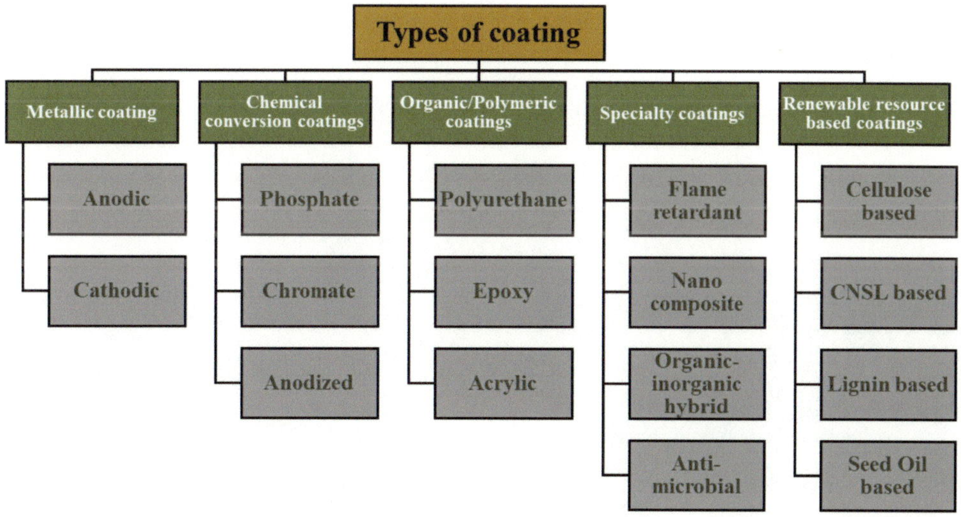

Figure 5.
Types of coatings used to combat corrosion.

layer between underlying metal and coating [22]. They are formed by immersing a metallic substrate in a chemical solution. Various types of conversion coatings are available. Phosphate coatings are produced on steels by dipping them in an appropriate phosphate solution. The thickness of these coatings depends upon the porosity of the coatings as it forms. These coatings increase corrosion resistance, absorb lubricant, promote adhesion, and enhance the appearance of the substrate. There are three types of phosphate coatings *a) Iron phosphates, b) zinc phosphate* and *c) manganese chromium phosphate* [21]. Chromate conversion coatings (CCCs) are generally formed by chemical or electrochemical treatment of metals and their alloys in a solution containing hexavalent chromium [Cr (VI)] and trivalent chromium [Cr(III)] ions with other components. These coatings form a complex chromate film over the entire surface of the metal. They are used on aluminum, magnesium, zinc, copper, cadmium, etc., [23]. Anodized coatings are formed by converting the workpiece of metal into an anode. This is usually done in order to form an oxide coating to increase the performance of the surface [24]. Polymeric coatings are widely applied for decorating as well protecting purposes. They act as a corrosion barrier between the underlying metal and corrosive media. These coatings consist of pigment, polymer, corrosion inhibitors, additives, etc., [25]. The protection provided by these coatings depends upon their ability to form highly resistant pathways between the cathodic and anodic areas on the surface of metal [26]. Acrylic, vinylic, epoxy, polyurethane, alkyd (oil-based) coatings are some of the examples of organic coatings used in corrosion protection [27]. With the advancement of technology, the need to introduce specialized coating with highly advanced functioning continues to increase. Various types of speciality coating are being investigated these days, such as Flame retardant coatings, nano-coatings, nanocomposite coatings, organic–inorganic hybrid coatings, etc. All these coatings are tailored according to their end-use and the type of environment they will be applied in [28].

Corrosion prevention coatings highly utilize petro-based products which are high cost, toxic, and constantly depleting. Constant research is being carried out to formulate better strategies that can meet the environmental and economical requirements. Renewable resources are environment friendly, less expensive, and naturally available. Renewable resource-based coatings are called *"green"*. **Figure 6** shows the different polymeric coating materials transformed from renewable resources.

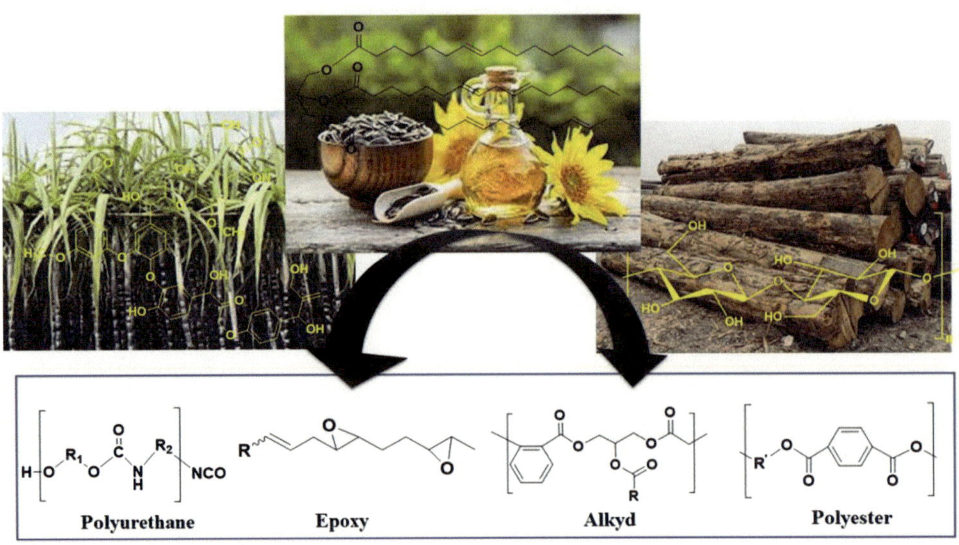

Figure 6.
Polymers based on various renewable resource.

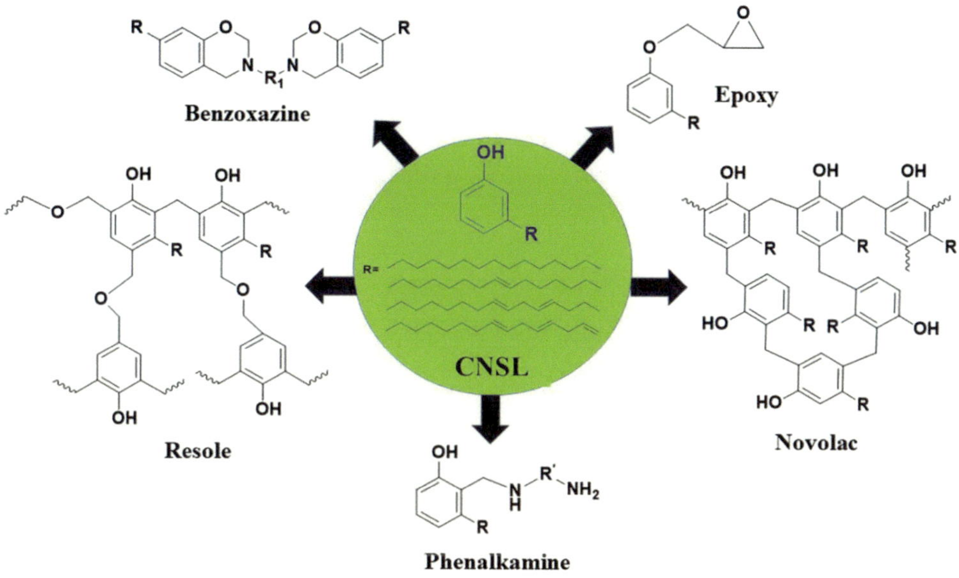

Figure 7.
CNSL derived polymers.

These green materials are highly employed in the field of corrosion protection in the form of corrosion-resistant coatings, inhibitors, pigments, composites, etc. Starch, cellulose, lignin, tannic acid, vegetable oils, and Cashew Nutshell Liquid (CNSL) are most commonly used for the synthesis of green coatings (**Figure** 7) [29].

Apart from using solely the polymeric coatings and relying on their chemical protective abilities, the formulations of nanostructures within the coating units have resulted in materials with higher corrosion protection efficiency [30–32]. These nanostructures can be formulated via., simple chemical reactions or by technologically advanced techniques like lithographical techniques [33]. The addition of inorganic or organic nanoparticles or units to improve corrosion protective performances has also been employed. Especially the inorganic nanoparticle dispersed within the polymeric matrix enhances the electrochemical stability, toughness, strength, and provides a tortuous pathway to the corrosive ions. The nanostructured surface coatings also tend to have higher surface hydrophobicity and scratch hardness [34]. Further, the *in-situ* synthesis of nanostructured components is also followed by many industries. The in-situ synthesis provides additional connectivity between the inorganic unit and organic unit, thereby generating well cross-linked high-density coatings [35, 36]. Due to all these value-added properties, nanocomposite coatings are getting more and more attention. In order to further enhance the efficacy of nanocomposite coatings, inhibitor-incorporated hybrid coatings are also considered innovative corrosion remediation systems [37].

2.2 Corrosion inhibitors

Pure metals and their alloys tend to react chemically and electrochemically with the corrosive environment to produce a stable compound. In this process, metals lose their mechanical strength and elasticity thereby becoming weak. Several strategies have been investigated which can retard or minimize or completely stop the cathodic or anodic or both reactions. Among them, utilization of corrosion inhibitors is a famous technique [7]. The chemical substances which when added in a small amount to the corrosive environment slow down/reduce the corrosion rate are called corrosion inhibitors. Inhibitors function through the adsorption of ions/molecules on the surface of the metal. They can drastically reduce the reaction rate either by decreasing/increasing the anodic and/or cathodic reactions or by decreasing the diffusion rate of reactants or decreasing the electrical resistance of the metallic surface. They can be easily applied and offer numerous advantages owing to their *in-situ* application without disturbing the process [38].

3. Corrosion testing methods

Two types of corrosion-resistant tests are generally used: *a)* salt spray test, *b)* Electrochemical impedance spectroscopy (EIS). Salt spray test (ASTM B117) is a standardized testing method conducted for the evaluation of the extent of corrosion resistance or protective coating. This test is generally carried out for 8-3000 hrs depending upon the coating in presence of 5% NaCl solution with pH between 6.5–7.2 [39]. EIS technique (ASTM G106) measures the impedance of the coating with the help of small amplitude, alternating current (AC). This AC signal is scanned at different frequencies to generate the spectrum for the electrochemical cell (the specimen) under the test [40].

4. Computational fitting and programming

Computational approaches like density functional theory (DFT), classical molecular dynamics (MD), Monte Carlo (MC) simulations, and others are becoming more and more preferred for corrosion studies. In order to improve the quality of the results, it is mostly combined with the experimental data. Particularly to verify the concept of corrosion inhibitors before performing the multiple experiments these studies were performed to find the thermodynamics, kinetics, energy levels (HOMO, LUMO, and Frontier molecular orbitals), bandgap energies, active sites, etc. All these theoretical calculation methodologies could be employed by artificial intelligence in the future to provide the platform which can create new and advanced corrosion inhibitors [41]. The complementary data with the experimental results puts the potential ground for the inhibitor to be explored further for corrosion inhibition applications. Even the small variations in the chemical structure which can result in better inhibition results can be evaluated through these computational approaches. Better pictorial view with the knowledge of possible interaction site, adsorption strength, and other quantum parameters proved to provide in-depth insight about the inhibition mechanism as done by different researchers in the area [42, 43].

5. Conclusion

Corrosion is an electrochemical process that causes metals to degrade at an increasing rate. Various types of corrosion can never be stopped completely but various methods can be applied to decrease the rate of corrosion. These methods include the utilization of coatings as well as corrosion inhibitors or a combination of which tends to protect metals from the attack of the corrosive environment thereby increasing the lifetime of metals. A lot of work is going on to build corrosion-resistant materials but more discussions over the topic and other advanced technologies which can improve the performance of the materials should be encouraged. The involvement of computational aided technologies can further help in reducing the search or designing of effective inhibitors without doing wet-lab experiments. The combined results of computational simulations and traditional electrochemical characterization techniques (impedance spectroscopy, variation in potential, corrosion current, rate of corrosion, and coating capacitance values) may further improve the understanding of this natural but unwanted phenomenon.

Author details

Shumaila Masood[1], Anujit Ghosal[2,3], Eram Sharmin[4], Fahmina Zafar[1*] and Nahid Nishat[1*]

1 Inorganic Material Research Laboratory, Department of Chemistry, Jamia Millia Islamia, New Delhi, India

2 Department of Food and Human Nutritional Sciences, The University of Manitoba, MB, Canada

3 Richardson Centre for Functional Foods and Nutraceuticals, The University of Manitoba, MB, Canada

4 Department of Pharmaceutical Chemistry, College of Pharmacy, Umm Al-Qura University, Makkah Al-Mukarramah, Saudi Arabia

*Address all correspondence to: fahmzafar@gmail.com; nishat_nchem08@yahoo.com

References

[1] Rodrigues R, Gaboreau S, Gance J, Ignatiadis I, Betelu S. Reinforced concrete structures: A review of corrosion mechanisms and advances in electrical methods for corrosion monitoring. Construction and Building Materials. 2021;**269**:121240. DOI: 10.1016/j.conbuildmat.2020.121240

[2] Ivaskova M, Kotes P, Brodnan M. Air pollution as an important factor in construction materials deterioration in Slovak Republic. Procedia Engineering. 2015;**108**:131-138. DOI: 10.1016/j.proeng.2015.06.128

[3] Prasad D. Corrosion and Natural Corrosion Inhibitors: A Case Study for C. microphyllus. In: Corrosion [Working Title]. London: IntechOpen; 2021. DOI: 10.5772/intechopen.100505

[4] Polder RB, Peelen WHA, Courage WMG. Non-traditional assessment and maintenance methods for aging concrete structures - technical and non-technical issues. Materials and Corrosion. 2012;**63**:1147-1153. DOI: 10.1002/maco.201206725

[5] Edison TNJI, Atchudan R, Pugazhendhi A, Lee YR, Sethuraman MG. Corrosion inhibition performance of spermidine on mild steel in acid media. Journal of Molecular Liquids. 2018;**264**:483-489. DOI: 10.1016/j.molliq.2018.05.087

[6] Ahmad Z. Types of corrosion. In: Principles of Corrosion Engineering and Corrosion Control. Amsterdam, Netherlands: Elsevier; 2006. pp. 120-270. DOI: 10.1016/B978-075065924-6/50005-2

[7] Raja PB, Ismail M, Ghoreishiamiri S, Mirza J, Ismail MC, Kakooei S, et al. Reviews on corrosion inhibitors: A short view. Chemical Engineering Communications. 2016;**203**:1145-1156. DOI: 10.1080/00986445.2016.1172485

[8] Sillanpää M, Shestakova M. Introduction. In: Electrochemical Water Treatment Methods. Amsterdam, Netherlands: Elsevier; 2017. pp. 1-46. DOI: 10.1016/B978-0-12-811462-9.00001-3

[9] Evans UR. The mechanism of the so-called "dry corrosion" of metals. Transactions of the Faraday Society. 1923; **19**:201-212. DOI: 10.1039/TF9231900201

[10] Frayne C. Environmental modification for cooling, heating and potable water systems. In: Shreir's Corrosion. Amsterdam, Netherlands: Elsevier; 2010. pp. 2930-2970. DOI: 10.1016/B978-044452787-5.00161-X

[11] Balan KP. Corrosion. In: Metallurgical Failure Analysis. Amsterdam, Netherlands: Elsevier; 2018. pp. 155-178. DOI: 10.1016/B978-0-12-814336-0.00009-3

[12] Laleh M, Hughes AE, Xu W, Haghdadi N, Wang K, Cizek P, et al. On the unusual intergranular corrosion resistance of 316L stainless steel additively manufactured by selective laser melting. Corrosion Science. 2019;**161**:108189. DOI: 10.1016/j.corsci.2019.108189

[13] Chen Y, Zhang W, Ding J, Yin P, Wang J, Zou Y. Research on water-line corrosion of carbon steel by wire beam electrode technique. MATEC Web of Conferences. 2016;**39**:01004. DOI: 10.1051/matecconf/20163901004

[14] Birbilis N, Hinton B. Corrosion and corrosion protection of aluminium. In: Fundamentals of Aluminium Metallurgy. Amsterdam, Netherlands: Elsevier; 2011. pp. 574-604. DOI: 10.1533/9780857090256.2.574

[15] Liu H, Gu T, Asif M, Zhang G, Liu H. The corrosion behavior and mechanism of carbon steel induced by extracellular polymeric substances of iron-oxidizing bacteria. Corrosion

Science. 2017;**114**:102-111. DOI: 10.1016/j.corsci.2016.10.025

[16] Hofacker S, Mechtel M, Mager M, Kraus H. Sol–gel: A new tool for coatings chemistry. Progress in Organic Coatings. 2002;**45**:159-164. DOI: 10.1016/S0300-9440(02)00045-0

[17] Artesani A, Di Turo F, Zucchelli M, Traviglia A. Recent advances in protective coatings for cultural heritage–an overview. Coatings. 2020; **10**:217. DOI: 10.3390/coatings10030217

[18] Khatoon H, Ahmad S. Vanadium pentoxide-enwrapped Polydiphenylamine/polyurethane nanocomposite: High-performance anticorrosive coating. ACS Applied Materials & Interfaces. 2019;**11**:2374-2385. DOI: 10.1021/acsami.8b17861

[19] Engan HE, Bløtekjær K. Acousto-optic interaction in few-mode optical fibers. In: Guided Wave Optical Components and Devices. Amsterdam, Netherlands: Elsevier; 2006. pp. 311-324. DOI: 10.1016/B978-012088481-0/50021-8

[20] Reghuraj AR, Saju KK. Black oxide conversion coating on metals: A review of coating techniques and adaptation for SAE 420A surgical grade stainless steel. Materials Today: Proceedings. 2017;**4**:9534-9541. DOI: 10.1016/j.matpr.2017.06.219

[21] Ahmad Z. Coatings. In: Principles of Corrosion Engineering and Corrosion Control. Amsterdam, Netherlands: Elsevier; 2006. pp. 382-437. DOI: 10.1016/B978-075065924-6/50008-8

[22] Chen X-B, Chong K, Abbott TB, Birbilis N, Easton MA. Biocompatible strontium-phosphate and manganese-phosphate conversion coatings for magnesium and its alloys. In: Surface Modification of Magnesium and Its Alloys for Biomedical Applications. Amsterdam, Netherlands: Elsevier; 2015. pp. 407-432. DOI: 10.1016/B978-1-78242-078-1.00015-3

[23] Pehkonen SO, Yuan S. Introduction and Background. Amsterdam, Netherlands: Elsevier; 2018. pp. 1-11. DOI: 10.1016/B978-0-12-813584-6.00001-6

[24] Ferreira MGS, Zheludkevich ML, Tedim J, Yasakau KA. Self-healing nanocoatings for corrosion control. In: Corrosion Protection and Control Using Nanomaterials. Amsterdam, Netherlands: Elsevier; 2012. pp. 213-263. DOI: 10.1533/9780857095800.2.213

[25] Taylor SR. The role of intrinsic defects in the protective behavior of organic coatings. In: Handbook of Environmental Degradation of Materials. Amsterdam, Netherlands: Elsevier; 2012. pp. 655-672. DOI: 10.1016/B978-1-4377-3455-3.00022-5

[26] Zanella C, Fedel M. Surface treatments on al alloys and composites. In: Encyclopedia of Materials: Metals and Alloys. Amsterdam, Netherlands: Elsevier; 2022. pp. 170-178. DOI: 10.1016/B978-0-12-819726-4.00116-2

[27] Popov BN. Organic coatings. In: Corrosion Engineering. Amsterdam, Netherlands: Elsevier; 2015. pp. 557-579. DOI: 10.1016/B978-0-444-62722-3.00013-6

[28] Radhamani A, Lau HC, Ramakrishna S. Nanocomposite coatings on steel for enhancing the corrosion resistance: A review. Journal of Composite Materials. 2020;**54**:681-701. DOI: 10.1177/0021998319857807

[29] Sharmin E, Ahmad S, Zafar F. Renewable resources in corrosion resistance. In: Corrosion Resistance. London: INTECH; 2012. DOI: 10.5772/31995

[30] Sharmin E, Akram D, Ghosal A, Rahman OU, Zafar F, Ahmad S. Preparation and characterization of

nanostructured biohybrid. Progress in Organic Coatings. 2011;**72**:469-472. DOI: 10.1016/j.porgcoat.2011.06.004

[31] Rawat NK, Ghosal A, Ahmad S. Influence of microwave irradiation on various properties of nanopolythiophene and their anticorrosive nanocomposite coatings. RSC Advances. 2014;**4**:50594-50605. DOI: 10.1039/C4RA06679K

[32] Zafar F, Ghosal A, Sharmin E, Chaturvedi R, Nishat N. A review on cleaner production of polymeric and nanocomposite coatings based on waterborne polyurethane dispersions from seed oils. Progress in Organic Coatings. 2019;**131**:259-275. DOI: 10.1016/j.porgcoat.2019.02.014

[33] Kuczyńska-Zemła D, Sotniczuk A, Pisarek M, Chlanda A, Garbacz H. Corrosion behavior of titanium modified by direct laser interference lithography. Surface and Coatings Technology. 2021;**418**:127219. DOI: 10.1016/j.surfcoat.2021.127219

[34] Ghosal A, Iqbal S, Ahmad S. NiO nanofiller dispersed hybrid soy epoxy anticorrosive coatings. Progress in Organic Coatings. 2019;**133**:61-76. DOI: 10.1016/j.porgcoat.2019.04.029

[35] Ghosal A, Rahman OU, Ahmad S. High-performance soya polyurethane networked silica hybrid nanocomposite coatings. Industrial & Engineering Chemistry Research. 2015;**54**:12770-12787. DOI: 10.1021/acs.iecr.5b02098

[36] Ghosal A, Ahmad S. High performance anti-corrosive epoxy–titania hybrid nanocomposite coatings. New Journal of Chemistry. 2017;**41**: 4599-4610. DOI: 10.1039/C6NJ03906E

[37] Rahimi A, Amiri S. Anticorrosion hybrid nanocomposite coatings with encapsulated organic corrosion inhibitors. Journal of Coatings Technology and Research. 2015;**12**:587-593. DOI: 10.1007/s11998-015-9657-4

[38] Raja PB, Sethuraman MG. Natural products as corrosion inhibitor for metals in corrosive media — A review. Materials Letters. 2008;**62**:113-116. DOI: 10.1016/j.matlet.2007.04.079

[39] Ito M, Ooi A, Tada E, Nishikata A. In situ evaluation of carbon steel corrosion under salt spray test by electrochemical impedance spectroscopy. Journal of The Electrochemical Society. 2020;**167**: 101508. DOI: 10.1149/1945-7111/ab9c85

[40] Simpson T, Moron P, Hampel H, Davis G, Shaw B, Arah C, et al. Electrochemical impedance measurements for evaluating and predicting the performance of organic coatings for atmospheric exposure. Corrosion Testing and Evaluation: Silver Anniversary Volume. 2009;**397**:16. DOI: 10.1520/STP39203S

[41] Dong C, Ji Y, Wei X, Xu A, Chen D, Li N, et al. Integrated computation of corrosion: Modelling, simulation and applications. Corrosion Communications. 2021;**2**:8-23. DOI: 10.1016/J.CORCOM.2021.07.001

[42] Hsissou R, Abbout S, Seghiri R, Rehioui M, Berisha A, Erramli H, et al. Evaluation of corrosion inhibition performance of phosphorus polymer for carbon steel in [1 M] HCl: Computational studies (DFT, MC and MD simulations). Journal of Materials Research and Technology. 2020;**9**:2691-2703. DOI: 10.1016/J.JMRT.2020.01.002

[43] Lgaz H, Saha SK, Chaouiki A, Bhat KS, Salghi R, Shubhalaxmi P, et al. Exploring the potential role of pyrazoline derivatives in corrosion inhibition of mild steel in hydrochloric acid solution: Insights from experimental and computational studies. Construction and Building Materials. 2020;**233**:117320. DOI: 10.1016/J.CONBUILDMAT.2019.117320

Chapter 2

Corrosion: Favoured, Yet Undesirable - Its Kinetics and Thermodynamics

Lukman O. Olasunkanmi

Abstract

This chapter describes the fundamentals of metal corrosion in relation to thermodynamics and kinetics. The chapter is so titled, because corrosion of metal is thermodynamically favourable. Moreover, it impacts negatively on economy and safety. Industries expend a substantial percentage of their budgets on corrosion control, and lose revenue due to corrosion damage. Effects of corrosion on industrial and public infrastructure cannot be overemphasized. Several accidents in the transportation and recreational industries have been linked to corrosion of metallic parts of respective gadgets. Some of these accidents are utterly catastrophic and fatal. Therefore, corrosion, albeit its thermodynamic favouability, is not desired by man. Metals corrode as a way of minimizing energy contents. Active metals are more stable in combined forms such as oxides, sulphides, and hydroxides, even though these forms are less useful to man. It appears the "price" to pay for extracting the pure forms of these metals from their ores is corrosion. This chapter presents fundamentals of thermodynamics and kinetics of metal corrosion, with emphasis on aqueous medium. It promises to serve as an introductory chapter for corrosion science students and as a concise material for tutors.

Keywords: corrosion, metallurgy, spontaneity, passivity, electrochemical

1. Introduction

Corrosion is a spontaneous disintegration of materials owing to their reactions with chemical constituents of the surroundings. Materials in this context may include metals, polymers and ceramics [1, 2]. However, corrosion is mostly used to refer to undesirable destruction of metals and alloys due to interactions with surrounding environment. The interactions or reactions as used here could be chemical or electrochemical in nature. More concisely, for the purpose of this chapter, corrosion would be described as chemical or electrochemical reaction between a metal and constituents of its environment. Emphasis will be placed on corrosion of metal in aqueous environments. Corrosion is a "favoured" process, requiring little or no energy input for its occurrence. Despite being naturally favoured, corrosion imposes a lot of economic strain, health and safety threat on human and society. It is a naturally favoured and an unavoidable process, yet, undesirable by man.

Corrosion is a way by which a metal assumes a low energy state, by combining with some other elements such as oxygen, hydrogen and sulphur. Metals are naturally not favoured to exist in free states. They are found in combined forms, often as ores. Pure metals are extracted from their respective ores with a lot of energy input. No wonder corrosion is also touted as "retro-extractive metallurgy" [3], as depicted in Eq. (1).

$$\text{Ore} \underset{\Delta G < 0}{\overset{\Delta G > 0}{\rightleftharpoons}} \text{Metal (free state)} \quad (1)$$

Extractive Metallurgy (forward); Corrosion (reverse)

Corrosion of metals is essentially an electrochemical process. A metal could assume an immune, active or a passive state when exposed to the environment, depending on the nature of the metal and the environment [1]. An index of the possible state of a metal in an environment is its electrochemical/redox potential (E_r). The half-cell equation for the reduction of a metal (M) could be written as:

$$M^{n+}{}_{(aq)} + ne^- \rightleftharpoons M_{(s)} \; E = E_r \text{ (Volts)} \quad (2)$$

Metals with highly positive E_r are naturally immune to oxidation, examples include Au and Pt. Those with negative E_r are active in the environment. Some of them may assume an active-passive state, depending on the nature of the environment and the properties of their corrosive products.

Understanding the thermodynamics and kinetics of metal corrosion would make corrosion science students to appreciate why corrosion occurs and how its rate could be mitigated. It would also help corrosion science tutors to lay a good foundation of the course for their students.

2. The fate of a metal in an environment

The behaviour of a metal upon exposure to an environment depends on the nature of the metal and the conditions of the environment. A metal exposed to a corrosive environment could behave in one of the following ways [1]:

i. Immune behaviour: the metal may be immune to an environment, such that it does not react in the environment. That is, the metal does not corrode. Such a metal is said to be thermodynamically stable in such an environment and would not undergo spontaneous corrosion. Metals with immune behaviour are very noble. They are un-reactive in virtually all environments. Examples include Au, Ag, Pt, and Pd.

ii. Active behaviour: a metal is said to be of active behaviour in an environment when it reacts with the environment and thereby corrodes. If the environment is an aqueous solution, such a metal dissolves in solution and forms soluble non-protective corrosion products. Since the corrosion product is soluble and non-protective, the dissolution of such a metal is continuous. A metal with active behaviour suffers significant weight loss in an aggressive solution. Examples include Fe, Al, Zn, Mg, Cu etc. Activity of metals may vary with environmental conditions such as corrosive ions, immersion time, temperature, etc. In a recent study, Dong *et al.* [4] compared the corrosion rates of Zn, Fe and Mg in simulated physiological solutions and found that Mg exhibited the highest corrosion rate. The study also illustrates how corrosion rate could vary with immersion time.

iii. Passive behaviour: a metal that exhibits passive behaviour might corrode at some instances but assume a non-responsive state over a period. Upon exposure to the corrosive environment, a metal with passive behaviour forms insoluble protective corrosion products, which cover the surface of the metal and shield it from further exposure to the corrosive environment. The protective film slows down further reaction with the environment. The films are usually more stable (less reactive) than the metals themselves. However, if the passive film is broken or dissolved, then the metal can revert to the active state, at which instance rapid dissolution of the metal could occur. In some cases, repassivation could follow the breakdown of passive film. Various factors could be responsible for passive film breakdown or instability. Among them are film thickness, nature of corrosive ions, pH, anodic potentials and so on. Feng *et al.* [5] observed that increased strain magnitude could also increase instability of passive film on carbon steel. A more detailed overview on the passivity of metals could be found in literature [6].

It is noteworthy that the behaviour of a metal depends on its "micro-environment". The natural or real practical environments are characterized by variable factors. The conditions of the environment may change with time. There could be a change in the cell pH, fluid flow rate, and temperature. Some reactions in the environmental might also result in solid deposits. The actual environment to which a metal responds is the immediate local environment at or near its surface. It is this micro-environment that determines the behaviour of the metal. Though some metals exhibit nearly universal behaviour irrespective of the environmental conditions. Metals such as Au, Pt and Ag typically exhibit noble or immune behaviour regardless of the environment, while metals like Na, K and Mg are generally active in nearly all aqueous environments. Metals likes titanium and tantalum assume passive state in a wide range of aqueous environments, though reactive in some other environments. Aluminium and zinc are very reactive metals and often exhibit active behaviour. However, they form stable protective passive films in some environments. The behaviour of such metals could be described as "active-passive" – active in some environments at some instances, and sometimes passive. The subsequent passivity is due to the change in the local environment at the surface of the metal, being occupied by the insoluble passive corrosion products.

Active-passive behaviour of iron in nitric acid was first observed in 1790 by Keir [7]. The thickness of passive films may vary with environmental conditions. Susceptibility of the film to breakdown also depends on the thickness [6]. Sato *et al.* [8] also reported that the composition and thickness of passive films on iron immersed a borate solution could change with change in potentials. Luo *et al.* [9] observed that alloy 59 (a Ni-Cr-Mo alloy with the least Fe content and highest Cr-Mo content) develops thicker passive film in air than sulphuric acid solution, and the constituents of the film vary in the two environments.

3. Corrosion: thermodynamic driving force

As pointed out earlier, corrosion of a metal is a spontaneous process. A metal in its pure state has a considerably higher energy than its corresponding ore. Metal corrodes in an attempt to minimize its energy, while assuming a more stable state.

Minerals or Ores	Constituent metal oxide, hydroxide or Sulphide	ΔH_f (kJ/mol)
Hematite	Fe_2O_3	−825.50
Bauxite	Al_2O_3	−1675.69
Zincite	ZnO	−350.46
Sphalerite	ZnS	−206.00
Brucite	$Mg(OH)_2$	−924.66
Chalcocite	Cu_2S	−79.50

Table 1.
Minerals or ores of some common metals, their chemical formulas and standard heats of formation.

Corrosion is therefore a means of energy minimization as a metal tends to return to its combined form in which it exists naturally. A basic illustration of this energy minimization is the exothermic nature of the formation of metal oxides, sulphides and hydroxides (**Table 1**). Formation of hematite (the world most important ore of iron) is accompanied by the release of a huge amount of energy (825.50 kJ/mol). Meanwhile, the relative energy of a free Fe atom is 0 kJ/mol. The same applies to other metals and their minerals.

3.1 Corrosion: electrochemical cell and electrode potential

Corrosion of metals is essentially an electrochemical process, involving both anodic oxidation and cathodic reduction reactions. A micro-electrochemical cell is established on the surface of a corroding metal. Perhaps for the sake of emphasis, electrochemical corrosion cell is a galvanic (or voltaic) cell. The progress of corrosion reaction is accompanied by flow of electric current (i), which has to do with movement of an electric charge across a potential difference. A corroding metal in an aqueous solution sets up a galvanic cell system comprising the metal (M) in contact with its metal ion (M^{n+}) such that an equilibrium is established. The site on the metal surface where dissolution of metal into its ion occurs is the anodic site. The cathodic site is set up not far from the anodic site. Each site constitutes a half-electrode reaction system, making up two half-cells, like that of a galvanic cell. The difference between the electrode potentials of the two half-cells can be expressed as:

$$E_{cell} = E_{cathode} - E_{anode} \tag{3}$$

In Eq. (3), both E_{cathode} and E_{anode} are reduction potentials. Since we are considering corrosion, the E_{cell} is equivalent to the corrosion potential, E_{corr}. For a spontaneous cell reaction as we have in corrosion, E_{anode} is always more negative than E_{cathode}, such that E_{cell} is always positive.

For a typical case of iron (Fe) corroding in an aerated aqueous solution as:

$$Fe_{(s)} + H_2O_{(l)} + ½\, O_2 \rightleftharpoons Fe(OH)_{2(g)} \tag{4}$$

The half-cell reactions equations can be expressed as:

$$\text{Anode}: Fe_{(s)} \rightleftharpoons Fe^{2+}{}_{(aq)} + 2\,e^- \quad E^0{}_{ox} = +0.409\ V \tag{5}$$

$$\text{Cathode}: ½\, O_{2(g)} + H_2O_{(l)} + 2\,e^- \rightleftharpoons 2\,OH^-{}_{(aq)} \quad E^0{}_{red} = +0.20\ V \tag{6}$$

In this case, E_{cell} = (0.2 + 0.409) V = 0.609 V. The more positive the E_{cell}, the more feasible the corrosion of the metal. Practically, the E_{cell} or E_{corr} is measured using a potentiostat. In such measurements, a half-cell must be chosen as a reference, e.g. standard hydrogen electrode (SHE). Electrode potential for SHE is set at 0 V (E_{H^+/H_2}). The electrode is made up of 1 M hydrogen ion (H^+) and hydrogen gas (H_2) at 1 atm, supported with a platinum plate. When coupled with an half-cell of unknown E, the cell potential recorded by the potentiometer is the electrode potential of the system with unknown E.

Potentiostats with three-electrode system are often used to measure electrochemical corrosion parameters. The three-electrode system consists of the working electrode, WE (the metal or alloy whose corrosion is being studied), reference electrode, RE (against which the corrosion potential of the metal/alloy is measured), and auxiliary/counter electrode, CE, which supports or protects the reference electrode against passage of current. Commonly used reference electrodes include calomel electrode, which composed of Hg/Hg_2Cl_2, sat'd. KCl, and silver-silver chloride electrode (Ag/AgCl, sat'd KCl). The latter is mostly used because of its relatively cheap cost and less toxicity compared to the mercury-based electrode.

Electrode potentials depend on concentrations of the species and temperature. Under the standard conditions of 25°C and 1 M concentration or 1 atm pressure of the species, it is referred to as the *standard electrode potential*. The standard electrode potentials (standard reduction potentials), E^0 for some species are listed in **Table 2**. The dependence of electrode potentials on concentrations of the species and temperature is expressed in the form of the Nernst equation, whose general form is:

$$E_r = E_r^0 - \frac{RT}{nF} \ln \left(\frac{[red]}{[oxi]} \right) \tag{7}$$

where n is the number of electrons transferred in the redox reaction, F is the Faraday constant, R is the gas constant and T is absolute temperature; [red] and [oxi] are the concentrations of the reduced and oxidized species, respectively.

3.2 Gibb's free energy and electrode potentials

Corrosion is characterized by lowering of Gibb's free energy or increasing electrochemical cell potential. For a corroding metal, a micro-electrochemical cell is created on the surface. The progress of metal corrosion is proportional to flow of current in the electrochemical cell. Thermodynamic parameters can be expressed for electrochemical systems. Since corrosion of metals is a constant pressure process, the Gibb's free energy (ΔG) is a good thermodynamic parameter for predicting its spontaneity. A spontaneous reaction is accompanied by energy minimization, which implies a negative ΔG, while a positive ΔG connotes non-spontaneous process. A non-spontaneous reaction requires energy input to proceed. For a system at equilibrium, $\Delta G = 0$. Corrosion being a spontaneous process has a negative ΔG. Thermodynamic favourability of metal corrosion could readily be predicted from the electrode potentials of the metal concerned. The change in Gibb's free energy (ΔG) per mole of an electrochemically reacting species is related to the electrode potentials as:

$$\Delta G = -nFE \tag{8}$$

where n is the valence of the species (number of electrons transferred), F is the Faraday's constant (1 F = 96,485 C) and E is the electrode potential (Volts). For a typical case of anodic dissolution of iron as it oxidizes from Fe to Fe^{2+} by losing two electrons ($Fe \rightleftharpoons Fe^{2+} + 2e^-$); $n = 2$.

Reduction half-reactions	E^0_{red} (V)
$O_3 + 2\,H^+ + 2\,e^- \rightleftharpoons O_2 + H_2O$	2.07
$H_2O_2 + 2\,H^+ + 2e^- \rightleftharpoons 2\,H_2O$	1.776
$Au^{3+} + 3\,e^- \rightleftharpoons Au$	1.68
$PbO_2 + 4\,H^+ + 2e^- \rightleftharpoons Pb^{2+} + 2\,H_2O$	1.467
$Cl_2 + 2\,e^- \rightleftharpoons 2\,Cl^-$	1.3583
$O_2 + 4\,H^+ + 4\,e^- \rightleftharpoons 2\,H_2O$	1.229
$Pt^{2+} + 2\,e^- \rightleftharpoons Pt$	1.20
$H_2O_2 + 2e^- \rightleftharpoons 2\,OH$-	0.88
$Hg_2^{2+} + 2\,e^- \rightleftharpoons Hg$	0.7961
$Cu^+ + e^- \rightleftharpoons Cu$	0.522
$O_2 + 2\,H_2O + 4\,e^- \rightleftharpoons 4\,OH^-$	0.401
$Cu^{2+} + 2e^- \rightleftharpoons Cu$	0.3402
$2\,H^+ + 2\,e^- \rightleftharpoons H_2$	0.000
$Fe^{3+} + 3\,e^- \rightleftharpoons Fe$	−0.036
$Pb^{2+} + 2\,e^- \rightleftharpoons Pb$	−0.1263
$Sn^{2+} + 2\,e^- \rightleftharpoons Sn$	−0.1364
$Ni^{2+} + 2\,e^- \rightleftharpoons Ni$	−0.23
$Co^{2+} + 2\,e^- \rightleftharpoons Co$	−0.28
$Fe^{2+} + 2\,e^- \rightleftharpoons Fe$	−0.409
$Cr^{3+} + 3\,e^- \rightleftharpoons Cr$	−0.74
$Zn^{2+} + 2\,e^- \rightleftharpoons Zn$	−0.7628
$Mn^{2+} + 2\,e^- \rightleftharpoons Mn$	−1.04
$Al^{3+} + 3\,e^- \rightleftharpoons Al$	−1.706
$Mg^{2+} + 2\,e^- \rightleftharpoons Mg$	−2.375
$Na^+ + e^- \rightleftharpoons Na$	−2.7109
$K^+ + e^- \rightleftharpoons K$	−2.924

Table 2.
Standard half-cell reduction potentials for reactions.

3.3 Impracticability of equilibrium electrochemical corrosion potentials

Corrosion involves both anodic and cathodic reactions. Each of these reactions is reversible and has associated electrode potential (E), which tends to attain an equilibrium value. At equilibrium, $\Delta G = E = 0$. However, attainment of this value is impracticable.

Practically, a bare (an oxide-free) metal surface releases metal ion into an aqueous solution (dissolution), leaving negatively charged electrons on the surface. This leads to an increase in the potential difference between the metal and the solution. The electrode potential becomes more negative. For the anodic dissolution of a metal, M, the half-cell reaction equation could be written as:

$$M \rightleftharpoons M^{n+} + ne^- \tag{9}$$

A more negative potential tends to retard dissolution but promotes deposition, according to Eq. (8) (i.e. ΔG is more positive for a more negative E). Since the process is reversible, continuous dissolution and/or deposition might lead to a stable (reversible) potential, E_r, which can be expressed in the form of Nernst Equation (for the reaction in Eq. (9)) as:

$$E_{r,M^{n+}/M} = E^0_{r,M^{n+}/M} + \frac{RT}{nF} ln\, a_{M^{n+}} \tag{10}$$

If a stable $E_{r,M^{n+}/M}$ is attained, dissolution would stop. However, in practice, $E_{r,M^{n+}/M}$ is never attained because electrons generated in Eq. (9) are always removed from the surface by the accompanied cathodic half-cell reaction, such as:

$$2H^+{}_{(aq)} + 2\,e^- \rightleftharpoons H_{2(g)} \tag{11}$$

Or

$$½\,O_{2(g)} + H_2O_{(l)} + 2\,e^- \rightleftharpoons 2\,OH^-{}_{(aq)} \tag{12}$$

in an acidic or basic medium respectively. The Nernst equation-type expressions for the reversible potential (E_r) for the cathodic reactions in Eqs. (11) and (12) respectively are:

$$E_{r,H^+/H_2} = E^0_{r,H^+/H_2} - \frac{RT}{F} ln \frac{P^{\frac{1}{2}}_{H_2}}{a_{H^+}} \tag{13}$$

$$E_{r,O_2/OH^-} = E^0_{r,O_2/OH^-} - \frac{RT}{4F} ln \frac{a^4_{OH^-}}{P_{O_2}} \tag{14}$$

where Eqs. (13) and (14) correspond to stoichiometrically adjusted forms of Eqs. (11) and (12) by multiplying the coefficients by ½ and 2 respectively for reduction of 1 mole of H^+ and O_2.

If E_r could be attained for reactions depicted by Eqs. (11) or (12), then E_r would be attained in Eq. (10). However, E_r in Eq. (13) or (14) is never stable due to continuous discharge of H_2 or consumption of O_2. Hence, attainment of a stable $E_{r,M^{n+}/M}$ is practically impossible and corrosion of metal, M is continuous.

3.4 Corrosion tendency based on electrochemical potentials and pH

Thermodynamics of electrochemical corrosion could be described as a function of electrode potential and hydrogen ions concentration (pH). This is often chatted as potential-pH diagram, popularly called the Pourbaix diagram, named after the original inventor. Pourbaix, a Belgium electrochemist and corrosion scientist invented the potential-pH diagram in 1963 for the description of thermodynamics of electrochemical corrosion. Pourbaix diagrams provide theoretical description of stability of a phase of metal/electrolyte system at a particular pH and potential. It is a kind of phase equilibrium diagram, though with different axes parameters compared to thermodynamics phase equilibrium diagram. Potential-pH diagram is often plotted at 25°C, 1 atm, and 10^{-6} M concentration of the ionic species. A typical Pourbaix plot comprises the redox potential on the vertical axis and the pH on the horizontal axis. Potential-pH diagram for iron in aqueous environment is shown in **Figure 1**. The diagram clearly shows the stable and passive regions for iron, based

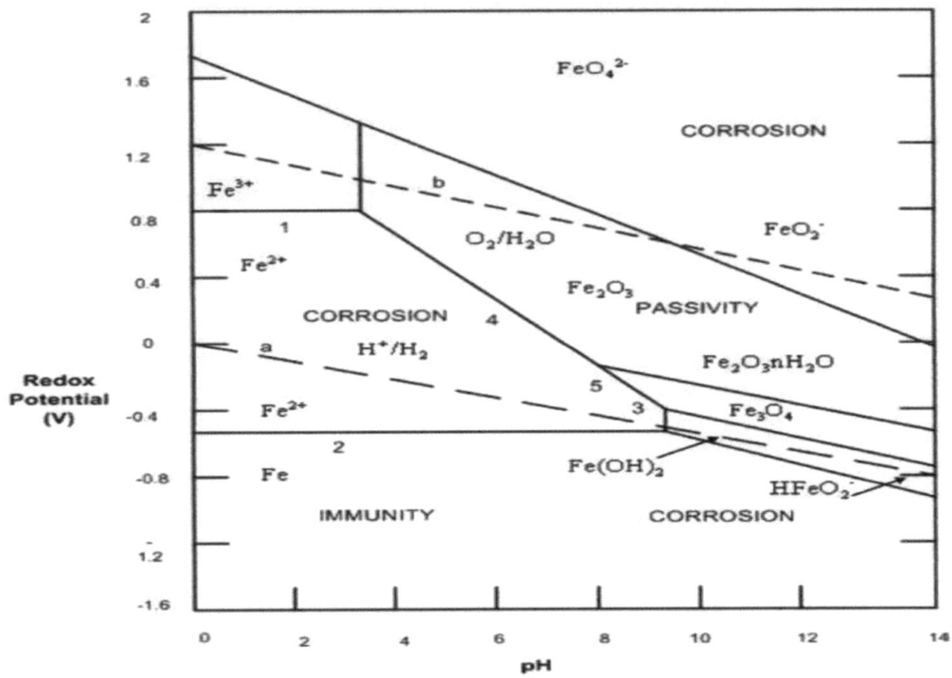

Figure 1.
Potential-pH diagram for iron in aqueous medium [10].

on the combination of potential and pH. At the very top of the diagram is corrosive region, where the potential is highly positive (above 1.8 V). At this region, iron would corrode at any combination of potential and pH. The region marked immunity at the lower portion of the diagram indicates the area where iron does not corrode. This region spans over a wide range of pH, and a limited range of highly negative potentials. Within this region, iron is immune to corrosion at various combinations of potentials and pH. In-between the two extreme top and bottom regions are regions where corrosion and/or passivity could occur, depending on the operating potentials and pH. The region marked "corrosion" between lines 1 and 2 covers a large area compared to the one at lower right-hand side. This implies that corrosion of iron at intermediate potentials between −500 mV and 1000 mV would progress more favourably in acidic pH than neutral and alkaline pH. At moderately positive potentials and neutral or alkaline pH, iron forms passive oxide film on the surface, which blocks further corrosion.

Pourbaix diagram could be used as a route to first principle corrosion simulation [11], a model for optimizing corrosiveness of a medium and designing materials with desired corrosion resistance. In a recent study, Nave and Kornev [12] constructed and applied 3D Pourbaix profile to establish the conditions for thermodynamic stability of tungsten-based compounds and describe the anodic dissolution of tungsten in aqueous solutions of potassium hydroxide. Beyond pure metals, Pourbaix profiles for multielement system such as Ni-Ti alloys having different ratios of the element have been proposed [13]. In an effort to overcome the challenges associated with developing Pourbaix profiles for complex compounds, a recent study by Patel *et al.* [14] introduced a more robust algorithm for modelling Pourbaix diagram for multicomponent materials.

4. Corrosion kinetics

Having discussed the propensity of metals to corrode, it is important to also highlight the rate at which metals corrode and mechanisms of corrosion reactions.

4.1 Corrosion rate

In a general term, corrosion rate (C_R) refers to the amount of metal loss to corrosion per unit time. The rate at which a metal corrodes can be monitored by various methods. This also determines the kind of C_R expression and units. Both the methods and expression (together with units) are in turn dependent on the technical system and type of corrosion being investigated. These methods may be classified as chemical, electrochemical, spectroscopic, and surface analysis based methods. Basically, corrosion rate (C_R) may be monitored by measuring any parameter that changes as corrosion reactions occur. For example:

- monitoring the change in weight of the metal upon corrosion (weight loss measurements);
- pressure of gas evolved (e.g. monitoring the H_2 gas evolution level);
- change in temperature upon corrosion reactions (thermometric measurement);
- spectroscopic analysis of amount of metal that goes into solution upon corrosion (e.g. AAS analysis of Fe in solution);
- amount of current that flows in the system (electrochemical measurements).

Each of these methods have associated merits and demerits. Corrosion rate may expressed as weight-loss of a metal per unit time per unit area according to the equation:

$$C_R = \frac{\Delta w}{At} \tag{15}$$

where Δw is the weight difference of the metallic block or plate at a set time interval (exposure time), t, and A is the exposed area of the metal. Being an analytical measurement, the ideal practice is to conduct repetitive measurements of Δw and utilize the average value in Eq. (15). This measurement is mainly applicable to general or uniform corrosion. However, it is the most used measurement of corrosion rate. For corrosion systems in which the metal/environment composition vary significantly with time and non-uniformity over the sites on corroding surface, measurement of C_R using Eq. (15) may be deficient. The results of such measurements must at least be supported with additional information such as the type of corrosion, dependence of corrosion rate on time and other relevant factors that prevail during the experiment. Eq. (15) is the simplest form of such expression, for which the units of Δw (g or nearest mg), A (nearest cm^2), and t (nearest h) would give the units of C_R as $gcm^{-2}\,h^{-1}$ or $mgcm^{-2}\,h^{-1}$. These units may not be applicable in reporting C_R in some other technical reports, for example, if C_R is to be expressed in the form of penetration rate (depth per unit time).

In an electrochemical experiment, corrosion rate could be measured as corrosion current density, in the form of corrosion current (mA) per unit area (cm^2) of

	mA cm^{-2}	mm year^{-1}	mpy	g m^{-2} day^{-1}
mA cm^{-2}	1	3.28 M/nd	129 M/nd	8.95 M/n
mm year^{-1}	0.306 nd/M	1	39.4	2.74 d
mpy	0.00777 nd/M	0.0254	1	0.0694 d
g m^{-2} day^{-1}	0.112 n/M	0.365 /d	14.4 /d	1

Table 3.
Conversion factors for various units of C_R [15].

corroding metal, i.e. mAcm^{-2}. Conversion factors from one C_R unit to another are listed in **Table 3.**
where:

- mpy = milli-inch per year
- n = number of electrons freed by the corrosion reaction
- M = atomic mass
- d = density

e.g.: 1 mA cm^{-2} = (3.28 M/nd) mm y^{-1} = (129 M/nd) mpy = (8.95 M/n) g m^{-2} day^{-1}

4.2 Factors affecting corrosion rate

Besides basic requirements for electrochemical corrosion to occur, which include the presence of anodic and cathodic sites, electrolyte and connectivity between the sites (to promote ionic conduction), there are secondary factors that affect corrosion rate. These factors are briefly discussed as follow.

- *Concentrations of dissolved oxygen (i.e. oxygen and oxidizers)*: corrosion rate and mechanism could be affected by the amount of dissolved oxygen. This is because, both anodic dissolution of metal and cathodic reactions are oxygen dependent. Generally, the higher the concentration of dissolved oxygen the higher the corrosion rate. Since anodic dissolution is an oxidation reaction, increased oxygen content in the microenvironment would increase the rate of metal oxidation. More so, when oxygen reduction predominates the cathodic reaction, dissolved oxygen increases the rate of cathodic reaction, which in turn speeds up the anodic reaction for charge balancing purpose. In other words, the rate of metal dissolution is directly proportional to the rate of oxygen reduction. A limited oxygen reduction or dissolved oxygen would lead to a decrease in metal dissolution and overall corrosion rate. Hydroxide and oxide deposits and protective film reduce the rate of oxygen diffusion onto the metal surface. They also tend to prevent conduction of metal ions from metal-oxide interface to oxide-liquid interface. Hence, they reduce corrosion rate, usually by passivation.

- *Concentration of dissolved salts:* dissolved salt increases conductivity of an electrolyte system. Since a requirement for corrosion to progress in ion mobility, increase in conductivity of the electrolyte due to dissolved salt increase corrosion rate. Generally, corrosion rates tend to increase when water conductivity increases. For this reason, corrosion rate is higher in saline water than freshwater. The corrosion rate in seawater is a function of numerous mutually dependent factors. However, according to Kirk and Pikul (1990) [16], if salinity exceeds 3%, water corrosivity decreases. This is due to decrease in oxygen solubility is in water with > 3% salinity, as posited by Weiss (1970) [17].

- *Temperature and pressure:* Generally, temperature increases the rate of metal corrosion, so does pressure. Just like every other electrochemical reactions, increase in temperature increases fluid flow and ionic mobility. Temperature can also affect scale formation and gas fugacity, which indirectly affect corrosion rate. Most corrosion models are accurate only within prescribed temperature ranges. For corrosive involving gases such as CO_2 and H_2S increase in operating pressure could increase the partial pressure of the corrosive gases. It should be noted that the reduction potential of the metal is dependent on fugacity of the gases present, according to the Nernst Equation. The dependence of corrosion rate (C_R) on temperature could be expressed in the form of Arrhenius equation as:

$$C_R = \text{Aex}(-E_a/\text{RT}) \tag{16}$$

where A is the pre-exponential factor, Ea is the activation energy, R is the gas constant and T is temperature.

- *Associated fluid dynamics*: increase in fluid flow rate increases the rate of metal loss. Fluid flow rate the reason for the particular type of corrosion, the "Flow-Accelerated Corrosion". Accelerated fluid drives corrosive ions to the reactive sites faster than a stagnant or unstirred solution. Fluid flow also has mechanical effects on passive films or deposits by driving them out of site, exposing the metal surface to direct attack of corrosive ions. Flow-accelerated corrosion is usually found at high flow rates around pipe/tube blockages, tube inlet ends, or in pump impellers.

- *Concentration of corrosive electrolyte*: generally the higher the concentration of corrosive electrolyte the higher the rate of metal corrosion. Though, change in concentration of corrosive electrolyte often lead to change in pH of the medium, which can affect corrosion rate and mechanism in different ways.

- *pH of the medium*: this affects the kind of reactions that take place at the cathode. For instance, in acidic pH, hydrogen ion reduction/hydrogen gas evolution predominates at the cathode. Oxygen reduction could also occur if the medium is aerated. Oxygen reduction is the primary cathodic reaction in neutral/basic solutions.

4.3 Corrosion mechanism

Corrosion, like many other chemical reactions usually involve more than one definable step. Interests often lie in the slowest step. Electrochemical corrosion involves release of ions and movement of electrons. Corrosion requires presence and movement of ions and electrons. A typical mechanism of corrosion of Fe in acidic medium is:

$$\text{Fe}_{(s)} \rightleftharpoons \text{Fe}^{2+}{}_{(aq)} + 2e^- \text{ (Anodic dissolution of metal)} \tag{17}$$

$$2\text{H}^+{}_{(aq)} + 2e^- \rightleftharpoons \text{H}_{2(g)} \text{ (Cathodic reaction–H}_2\text{ gas evolution)} \tag{18}$$

The process is often more complex such that, metal ions may go into solution as complexes or even; precipitate as hydroxides, oxides, sulphides etc.

Electrochemical corrosion mechanism of an active metal in an aqueous environment can be expressed generally as [18]:

$$M_{(s)} \rightleftharpoons M^{m+}{}_{(aq)} + me^- \quad (19)$$

$$mH^+{}_{(aq)} + me^- \rightleftharpoons m/2\ H_{2(g)} \quad (20)$$

$$\text{Overall}: M_{(s)} + mH^+{}_{(aq)} \rightleftharpoons M^{m+}{}_{(aq)} + m/2\ H_{2(g)} \quad (21)$$

In a deaerated acidic medium. Or.

$$M_{(s)} \rightleftharpoons M^{m+}{}_{(aq)} + me^- \quad (22)$$

$$mH_2O_{(l)} + me^- \rightleftharpoons mOH^-{}_{(aq)} + m/2\ H_{2(g)} \quad (23)$$

$$\text{Overall}: M_{(s)} + mH_2O_{(l)} \rightleftharpoons M^{m+}{}_{(aq)} + mOH^-{}_{(aq)} + m/2\ H_{2(g)} \quad (24)$$

In a deaerated neutral/basic medium.

In an aerated environment, oxygen plays a prominent role in the reaction and the mechanisms look like [18]:

$$M_{(s)} \rightleftharpoons M^{m+} + m\ e^- \quad (25)$$

$$m/4\ O_{2(g)} + m\ H^+{}_{(aq)} + m\ e^- \rightleftharpoons m/2\ H_2O \quad (26)$$

$$\text{Overall}: M_{(s)} + m/4\ O_{2(g)} + m\ H^+{}_{(aq)} \rightleftharpoons M^{m+} + m/2\ H_2O_{(l)} \quad (27)$$

In an aerated acidic environment.

Or

$$M_{(s)} \rightleftharpoons M^{m+} + m\ e^- \quad (28)$$

$$m/4\ O_{2(g)} + m/2\ H_2O_{(l)} + m\ e^- \rightleftharpoons m\ OH^- \quad (29)$$

Overall: $M_{(s)} + m/4\ O_{2(g)} + m/2\ H_2O_{(l)} \rightleftharpoons M^{m+} + m\ OH^-{}_{(aq)}$.
In an aerated neutral or basic environment.

5. Corrosion control methods

As contained in the previous sections, whether a material would corrode or not depends on a number of factors. The extent and rate of corrosion also varies, depending on the nature of the metal or alloy, corrosive medium, pH, temperature and so on. While it may be difficult to “work against nature”, to completely stop metal corrosion, several methods of abating corrosion damage have been identified. Most of these methods reduce corrosion rate, rather than “inverting” the thermodynamics. Although, thermodynamic susceptibility might be influenced in some instances. Choosing a suitable corrosion control method requires proper understanding of corrosion type and mechanism. No particular method had been adjudged universally effective to mitigate all forms of corrosion. Common corrosion control methods are highlighted below.

- *Material selection:* this method entails careful selection of metal or alloy that is immune to corrosion in an environment. While this sounds as a good idea, such materials do not often possess the desired mechanical properties for prospective technological design. Besides, using such materials for engineering purposes could be very expensive. In other words, striking a balance between material cost, mechanical properties and corrosion resistance is not a straightforward task.

- *Engineering design:* careful structural design might reduce or eliminate certain forms of corrosion. A design that eliminate accumulation of water on metal surface might reduce the rate of top-of-line corrosion (TLC), which is greatly associated with water accumulation [19]. Crevice corrosion could be minimized if engineering design minimizes stagnation of fluids and adopts welded rather than bolted joint. Design that eliminates or reduces turbulence flow could reduce the rate of erosion corrosion [20]. The major limitation to using engineering design to control corrosion is that the choice of a design might not meet other important criteria of the structure, aside reducing corrosion propensity of the material. A wrong design might be costly and yet not suitable for optimum application of the structure [21, 22].

- *Cathodic and anodic protection*: electrochemical potentials of a metal could be modified to suppress cathodic or anodic corrosion by shifting the potential cathodically or anodically, respectively. This could be achieved via impressed current system [23]. Cathodic shift increases the immunity of the metal by reducing electron uptake for the cathodic reaction [24]. A more positive or anodic shift of the potential increases the passivity behaviour of the metal [22, 25]. The use of sacrificial materials known as sacrificial anode/cathode is also a known practice. A more active metal (sacrificial anode) could be used to protect a metal of interest. Parthiban *et al.* [26] reported the use of magnesium alloy anode for the cathodic protection of steel embedded in concrete. It was observed that the mechanism of cathodic protection with the sacrificial anode could be correlated to the removal of corrosive ions such as chloride from the vicinity of steel.

- *Protective coatings:* organic or inorganic coatings are often used to protect the surface of active metals. Such coatings could serve as barrier protection and/or chemical inhibition. Besides corrosion mitigation, special coating materials could provide aesthetic functions, improving the surface appearance of the metal [27]. An extensive review on protective coating on magnesium alloy is available in literature [28].

- *Corrosion inhibitors:* chemical substances could be added to corrosive solution in trace quantities to reduce corrosion rate of metal. These additives, which often reduce the rate of metal corrosion by adsorbing as thin film on metallic surface are called corrosion inhibitors. Several corrosion inhibitors have been designed. Notable among them are benzotriazole (a popular inhibitor of copper corrosion) [29], quinoxalines [30–32], benzimidazoles [33], pyrimidines [34, 35], pyridazines [36], hydantoins [37], ionic liquids [38] and so on. Most organic corrosion inhibitors contain heteroatoms and pi-electron systems that aid adsorption of their molecules onto active sites on the metal surface. There are also inorganic inhibitors such as salts of zinc, copper, nickel, and arsenic. Earlier efficient inorganic corrosion inhibitors are chromate based, but their use have been discouraged due to their high toxicity.

Author details

Lukman O. Olasunkanmi
Department of Chemistry, Faculty of Science, Obafemi Awolowo University, Ile-Ife, Nigeria

*Address all correspondence to: waleolasunkanmi@gmail.com

References

[1] Davis, J.R., Corrosion: Understanding the basics, ASM International, 2000.

[2] Terminology relating to corrosion and corrosion testing, American Society for Testing and Materials Designation G 15-99b (Revised), 2000.

[3] Fontana, M.G., Corrosion Engineering, New York, McGraw-Hill, 1986.

[4] Dong, H., Lin, F., Boccaccini, A.R. and Virtanen, S. (2021). Corrosion behavior of biodegradable metals in two different simulated physiological solutions: Comparison of Mg, Zn and Fe. *Corrosion Science*, 182, 109278.

[5] Feng, X., Zuo, Y., Tang, Y., Zhao, X. and Zhao, J. (2012). The influence of strain on the passive behavior of carbon steel in cement extract. *Corrosion Science,* 65, 542–548.

[6] Sato N. (1990). An overview of the passivity of metals. *Corrosion Science*, 3 (I), 1–19.

[7] Keir, J. (1790). Experiments and observations on the dissolution of metals in acids and their precipitations; with an account of a new compound acid menftruum, useful in fume technical operations of parting metals. *Phil. Trans.*, 80, 359.

[8] Sato, N., Kudo, K. and Nishimura, R. (1976). Depth Analysis of Passive Films on Iron in Neutral Borate Solution. *J. Electrochem. Soc.,* **123** 1419.

[9] Luo, H., Gao, S., Dong, C., Li, X. (2014). Characterization of electrochemical and passive behaviour of Alloy 59 in acid solution. *Electrochimica Acta*, 135, 412–419.

[10] Ahmad, Z. (2006). Basic concepts in corrosion, in principles of corrosion engineering and corrosion control, 9–56, https://doi.org/10.1016/B978-075065924-6/50003-9.

[11] Perry, S.C., Gateman, S.M., Stephens, L.I., Lacasse, R., Schulz, R. and Mauzeroll, J. (2019). Pourbaix diagrams as a simple route to first principles corrosion simulation *Journal of The Electrochemical Society*, 166 (11) C3186-C3192.

[12] Nave, M.I. and Kornev, K.G. (2017). Complexity of products of tungsten corrosion: comparison of the 3D Pourbaix diagrams with the experimental data. *Metallurgical and Materials Transactions A*, 48A, 1414.

[13] Ding, R., Shang, J.X., Wang, F.H., Chen. Y. (2018). Electrochemical Pourbaix diagrams of NiATi alloys from first-principles calculations and experimental aqueous states. *Computational Materials Science,* 143, 431–438.

[14] Patel, A.M., Nørskov, J.K., Persson, K.A. and Montoya, J.H. (2019). Efficient Pourbaix diagrams of many-element compounds. *Phys. Chem. Chem. Phys.*, 21, 25323

[15] https://www.corrosion-doctors.org/Principles/Conversion.htm. Retrieved on 22nd April, 2021.

[16] Kirk, W. W., and Pikul, S. J. (1990). Seawater corrosivity around the world: results from three years of testing. In: C. H. Baloun (Ed.), Corrosion in Natural Waters (pp. 2–36). Philadelphia: American Society for Testing and Materials.

[17] Weiss R.F. The solubility of nitrogen, oxygen and argon in water and seawater. Deep Sea Research and Oceanographic Abstracts. 1970;17(4): 721–735. doi: 10.1016/0011-7471(70)90037-9

[18] Stanbury E. E. and Buchanan R.A. (2000). Fundamentals of Electrochemical Corrosion: *ASM International*, p 22 http://site.ebrary.com/id/10320341?ppg=22

[19] Larrey, D. and Gunaltun, Y.M. (2000). Correlation of cases of top of line corrosion with calculated water condensation rates, NACE International.

[20] Vyas, B. and Hansson, I.L. (1990). The cavitation erosion-corrosion of stainless steel. *Corrosion science*, 30, 761–770.

[21] Locke, C.E. (1987). Anodic protection. In: ASM handbook. *ASM International*, 13, 463–465.

[22] Riggs, O.L. and Locke, C.E. (1981). Anodic protection: theory and practice in the prevention of corrosion. Plenum Press.

[23] C. Christodoulou, C., Glass, G., Webb, J., Austin, S., and Goodier, C. (2010). Assessing the long term benefits of Impressed Current Cathodic Protection. *Corrosion Science* 52, 2671–2679.

[24] Morgan, J.H. (1987). Cathodic protection. 2nd Ed., NACE International

[25] Benedict, R.I. (1986). Anode resistance fundamentals and applications – classic paper and reviews. NACE International.

[26] Parthiban, G.T., Parthiban, T., Ravi, R., Saraswathy, V., Palaniswamy, N. and Sivan, V. (2008). Cathodic protection of steel in concrete using magnesium alloy anode *Corrosion Science,* 50, 3329–3335.

[27] Edwards, J. (1997). Coating and surface treatment systems for metals: A comprehensive guide to selection. Finishing Publication Ltd. and ASM International.

[28] Gray, J.E. and Luan, B. (2002). Protective coatings on magnesium and its alloys — a critical review. *Journal of Alloys and Compounds*, 336 (1–2), 88–113.

[29] Walker, R. (1970), "The use of benzotriazole as a corrosion inhibitor for copper", *Anti-Corrosion Methods and Materials*, 17 (9), 9–15. https://0-doi-org.ujlink.uj.ac.za/10.1108/eb006791.

[30] Olasunkanmi, L.O., Kabanda, M.M. and Ebenso, E.E. (2016). Quinoxaline derivatives as corrosion inhibitors for mild steel in hydrochloric acid medium: Electrochemical and quantum chemical studies. *Physica E*, 76, 109–126.

[31] Olasunkanmi, L.O., Obot, I.B., Kabanda, M.M. and Ebenso, E.E. (2015). Some Quinoxalin-6-yl Derivatives as Corrosion Inhibitors for Mild Steel in Hydrochloric Acid: Experimental and Theoretical Studies. *J. Phys. Chem. C,* 119, 16004–16019.

[32] Olasunkanmi, L.O. and Ebenso, E.E. (2020). Experimental and computational studies on propanone derivatives of quinoxalin-6-yl-4,5-dihydropyrazole as inhibitors of mild steel corrosion in hydrochloric acid. *Journal of Colloid and Interface Science,* 561, 104–116.

[33] Yadav, M., Kumar, S., Purkait, T., Olasunkanmi, L.O., Bahadur, I. and Ebenso, E.E. (2016). Electrochemical, thermodynamic and quantum chemical studies of synthesized benzimidazole derivatives as corrosion inhibitors for N80 steel in hydrochloric acid. *Journal of Molecular Liquids,* 213, 122–138.

[34] Verma, C., Olasunkanmi, L.O., Ebenso, E.E. and Quraishi, M.A. (2018). Adsorption characteristics of green 5-arylaminomethylene pyrimidine-2,4,6-triones on mild steel surface in acidic medium: Experimental and computational approach. *Results in Physics*, 8, 657–670.

[35] Verma, C., Olasunkanmi, L.O., Ebenso, E.E. and Quraishi, M.A. (2016). Adsorption behavior of glucosamine-based, pyrimidine-fused heterocycles as green corrosion inhibitors for mild steel: experimental and theoretical studies. *J. Phys. Chem. C*, 120, 11598–11611.

[36] Olasunkanmi, L.O., Mashuga, M.E. and Ebenso, E.E. (2018). Surface protection activities of some 6-substituted 3-chloropyridazine derivatives for mild steel in 1 M hydrochloric acid: Experimental and theoretical studies. *Surfaces and Interfaces*, 12, 8–19.

[37] Olasunkanmi, L.O., Moloto, B.P., Obot, I.B., and Ebenso, E.E. (2018). Anticorrosion studies of some hydantoin derivatives for mild steel in 0.5 M HCl solution: Experimental, quantum chemical, Monte Carlo simulations and QSAR studies Journal of Molecular Liquids 252, 62–74.

[38] Yesudass, S., Olasunkanmi, L.O., Bahadur, I., Kabanda, M.M., Obot, I.B. and Ebenso, E.E. (2016). Experimental and theoretical studies on some selected ionic liquids with different cations/anions as corrosion inhibitors for mild steel in acidic medium. *Journal of the Taiwan Institute of Chemical Engineers*, 64,252–268.

Section 2

Corrosion Inhibitors

Chapter 3

Copper, Iron, and Aluminium Electrochemical Corrosion Rate Dependence on Temperature

Mykhaylo Viktorovych Yarmolenko

Abstract

Our investigations show that electrochemical corrosion of copper is faster than electrochemical corrosion of aluminium at temperatures below 100°C. Literature data analysis shows that the Al atoms diffuse faster than the Cu atoms at temperatures higher than 475°C, Al-rich intermetallic compounds (IMCs) are formed faster in the Cu-Al system, and the Kirkendall plane shifts towards the Al side. Electrochemical corrosion occurs due to electric current and diffusion. An electronic device working time, for example, depends on the initial copper cover thickness on the aluminium wire, connected to the electronic device, temperature, and volume and dislocation pipe diffusion coefficients, so copper, iron, and aluminium electrochemical corrosion rates are investigated experimentally at room temperature and at temperature 100°C. Intrinsic diffusivities ratios of copper and aluminium at different temperatures and diffusion activation energies in the Cu-Al system are calculated by the proposed methods here using literature experimental data. Dislocation pipe and volume diffusion activation energies of pure iron are calculated separately by earlier proposed methods using literature experimental data. Aluminium dissolved into NaCl solution as the Al^{3+} ions at room temperature and at temperature 100°C, iron dissolved into NaCl solution as the Fe^{2+} (not Fe^{3+}) ions at room temperature and at temperature 100°C, copper dissolved into NaCl solution as the Cu^{+} ions at room temperature, and as the Cu^{+} and the Cu^{2+} ions at temperature 100°C. It is found experimentally that copper corrosion is higher than aluminium corrosion, and the ratio of electrochemical corrosion rates, $k_{Cu}/k_{Al} > 1$, decreases with temperature increasing, although iron electrochemical corrosion rate does not depend on temperature below 100°C. It is obvious because the melting point of iron is higher than the melting point of copper or aluminium. It is calculated that copper electrochemical corrosion rate is approximately equal to aluminium electrochemical corrosion at a temperature of about 300°C, so the copper can dissolve into NaCl solution mostly as the Cu^{2+} ions at a temperature of about 300°C. The ratio of intrinsic diffusivities, $D_{Cu}/D_{Al} < 1$, increases with temperature increasing, and intrinsic diffusivity of aluminium could be approximately equal to intrinsic diffusivity of copper at a temperature of about 460°C.

Keywords: electrochemical corrosion, metallic coatings, electrolysis, diffusion, intermetallic compounds, phases formation kinetics, copper, aluminium, iron, Kirkendall-Frenkel porosity, Kirkendall shift, activation energy

1. Introduction

An Al wire coated with a thin Cu cover (≈15-μm thickness), utilised near an automobile motor, is heated to temperatures about 373–473 K (100–200°C). Intermetallics (IMCs) can be formed at the Cu/Al interface and grow gradually during heating at such temperatures. The IMC layers are brittle and have high resistivity. Therefore, for assurance of the reliability of the product, information on the growth behaviour of the IMC layers during heating is essentially important [1]. **Figure 1** shows the problem: an electronic device working time, t_0, depends on initial Cu cover thickness, X_{Cu}, and temperature. The electric conductivity of copper is higher than the electric conductivity of aluminium in approximately two times, but the formation of intermetallic phases induces a significant increase in contact resistance, which is found to increase linearly with the thickness of the intermetallics formed [2]. The temperature range used to produce the intermetallic phases was from 250 to 515°C. Moreover, the presence of an electrical field greatly accelerated the kinetics of formation of intermetallic phases and altered significantly their morphology, and the impaired mechanical integrity of the Al-Cu bimetallic joints treated by an electrical current was clearly demonstrated by an extensive cracking not only across the whole intermetallic bandwidth but also within different phases and at a neighbouring interface [2]. Three-phase thickness, X_{123}, can be estimated in such a way. The mass conservation law gives

$$X_{Cu}(t=0)\cdot 1 = \frac{9}{9+4}X_3(t_0) + \frac{1}{1+1}X_2(t_0) + \frac{1}{1+2}X_3(t_0) \approx \frac{1.526X_{123}}{3} \approx 0.509X_{123}, \text{and } X_{123} \approx 2X_{Cu}, \quad (1)$$

so three-phase general thickness is approximately greater in two times than the initial Cu cover thickness.

Otherwise, it was proved experimentally that a thin Al pad (≈1-μm thickness) can prevent gold and copper corrosion because the intermetallics formation rate in Au-Al system is much higher than the intermetallic formation rate in Cu-Al system, so it is possible to use Cu instead of Au for wire bonding in microelectronic packaging, and Cu has higher electric conductivity, higher thermal conduction, and

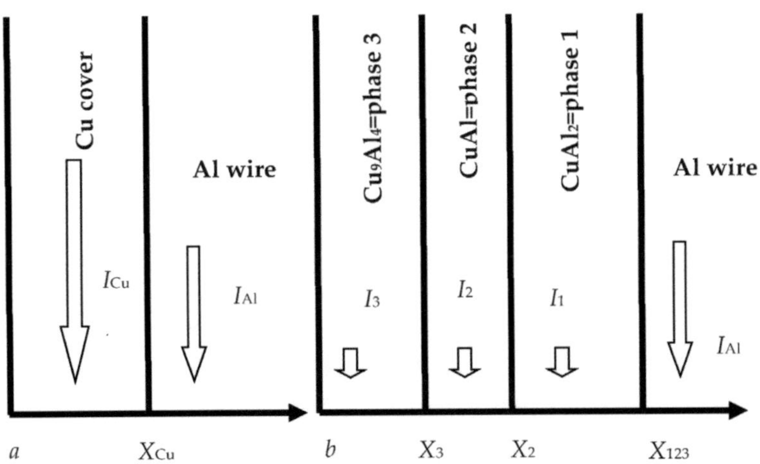

Figure 1.
(a) Initial stage (t = 0): An electronic device is working, since the electric current, $I_a = I_{Cu} + I_{Al}$, has optimal value; (b) final stage (t = t_0): The electronic device is not working, because the electric current, $I_b = I_3 + I_2 + I_1 + I_{Al}$, has too small value, since pure Cu cover has disappeared.

lower material cost than Au [3]. Corrosion and intermetallic rate formation in gold and copper wire bonding in microelectronics packaging were investigated in [3] at temperatures T_1 = 175°C, T_2 = 200°C, and T_3 = 225°C during 120, 240, 360, and 480 h. The authors have reported that cross-sectional analysis of the Cu ball on Al pad confirmed that corrosion occurred at temperatures about T = 200°C primarily beneath the Cu balls and did not initiate from the Al pad, formation of $CuCl_2$ did not allow self-passivation of Cu to occur, so the rate of copper corrosion increased, and the rate of Cu-Al intermetallics formation was found to be three to five times slower than Au-Al intermetallic formation at all three annealing temperatures. So, copper dissolved into NaCl solution as Cu^{2+} ions at temperatures about T = 200°C, as we expected. They did not investigate corrosion rate dependence of copper and aluminium on temperature. Moreover, phase layers general thicknesses for Cu-Al system were calculated [3]:

$$X_{123}^2 = K_{123}t + K_{01} = K_0 e^{-Q/(RT)} + K_{01} = 3.52 \cdot 10^{-4} \mu m^2/s \cdot e^{-25500 Jmol^{-1}/(RT)} t + 0.44 \mu m^2, \quad (2)$$

where $R \approx 8.314$ JK^{-1} is the gas constant, and K_{01} is the constant related to initial IMC thickness. General reaction rates of IMC formation were calculated: $K_{123}(T_1)$ = $3.57 \cdot 10^{-7}$ μm²/s, $K_{123}(T_2)$ = $6.26 \cdot 10^{-7}$ μm²/s, and $K_{123}(T_3)$ = $7.15 \cdot 10^{-7}$ μm²/s. The pre-exponential factor and IMC formation activation energy was calculated: $K_0 \approx 3.52 \cdot 10^{-4}$ μm²/s, $Q \approx 25.5$ kJ/mol. We can use these results to calculate the electronic device working time by Eq. (14) in [4] at different temperatures:

$$t_0 \approx \frac{X_{Cu}^2}{C_3^2 K_{123}} = \frac{169}{81} \frac{X_{Cu}^2}{K_{123}} \approx \frac{2X_{Cu}^2}{K_0} e^{Q/(RT)} \approx 5900 \cdot X_{Cu}^2 \left[\mu m^2\right] \cdot e^{25.5 kJmol^{-1}/(RT)} s; \quad (3)$$

$$t_0(T_1 = 175^oC = 448K) \approx 5900 \cdot 225 \cdot e^{25500/(8.314 \cdot 448)} s \approx 40 years;$$

$$t_0(T_2 = 200^oC) \approx 28 years;$$

$$t_0(T_3 = 225^oC) \approx 21 years; t_0(T_4 = 300^oC) \approx 9 years; t_0(T_5 = 350^oC) \approx 6 years.$$

Other researchers have obtained [5]: $K_{123}(T_4$ = 300°C) = $4.2 \cdot 10^{-4}$ μm²/s, $K_{123}(T_5$ = 350°C) = $3.4 \cdot 10^{-3}$ μm²/s. Eq. (3) gives

$$t_0(T_4 = 300^oC) \approx \frac{2X_{Cu}^2}{K_{123}} \approx 12 days; t_0(T_5 = 350^oC) \approx \frac{2X_{Cu}^2}{K_{123}} \approx 1.5 days.$$

We can calculate: $Q \approx 124$ kJ/mol; $K_0 \approx 8.5 \cdot 10^{-5}$ m²/s = $8.5 \cdot 10^7$ μm²/s;

$$t_0(T = 175^oC) \approx 5.3 \cdot 10^{-6} \cdot e^{124 Jmol^{-1}/(448R)} s \approx 49 years; t_0(T = 200^oC) \approx 8.4 years;$$

$$t_0(T = 225^oC) \approx \frac{2X_{Cu}^2}{K_0} e^{Q/(RT)} \approx 5.3 \cdot 10^{-6} \cdot e^{124 Jmol^{-1}/(498R)} s \approx 1.7 years.$$

It was reported in [1] that the growth of layer 1 is controlled predominantly by boundary diffusion, but that of layers 2 and 3 are governed mainly by volume diffusion at temperatures T = 483–543 K (210–270°C) for various periods up to 3.456Ms (960 h). The authors obtained: $K_{01} \approx 5.3 \cdot 10^{-7}$ m²/s, $Q_1 \approx 86$ kJ/mol, $K_{023} \approx 4.2 \cdot 10^{-5}$ m²/s, $Q_{23} \approx 146$ kJ/mol. We can calculate: K_{123} (T_6 = 210°C) = $1.5 \cdot 10^{-6}$ μm²/s,

$$t_0(T = 210^oC) \approx \frac{2X_{Cu}^2}{K_{123}} \approx 9.6 years.$$

The less temperature is, the higher contribution of grain-boundary diffusion and dislocation pipe diffusion to the layers growth is, so models of grain-boundary diffusion and dislocation pipe diffusion involving outflow into volume should be taken into account [6–8].

Diffusion activation energy of Al is less than diffusion activation energy of Cu ($Q_{Al} < Q_{Cu}$) at temperatures from 160–250°C for mutual diffusion in copper-a luminium thin film double layers, but the pre-exponential factors are different in 10 times [9]:

$$D_{Al}^{*} = 4 \cdot 10^{-5} e^{-121kJmol^{-1}/(RT)} m^2/s, D_{Cu}^{*} = 9.5 \cdot 10^{-4} e^{-135kJmol^{-1}/(RT)} m^2/s, \quad (4)$$

in θ-phase (phase 1) $CuAl_2$, $C_{Al} = 2/3 \approx 0.67$, $C_{Cu} = 1/3 \approx 0.33$;

$$D_{Al}^{*} = 1.5 \cdot 10^{-11} e^{-68kJmol^{-1}/(RT)} m^2/s, D_{Cu}^{*} = 1 \cdot 10^{-6} e^{-106kJmol^{-1}/(RT)} m^2/s, \quad (5)$$

in η_2-phase (phase 2) CuAl, $C_{Al} = C_{Cu} = 1/2 = 0.5$;

$$D_{Al}^{*} = 1.7 \cdot 10^{-7} e^{-116kJmol^{-1}/(RT)} m^2/s, D_{Cu}^{*} = 2.4 \cdot 10^{-6} e^{-125kJmol^{-1}/(RT)} m^2/s, \quad (6)$$

in γ_2-phase (phase 3) Cu_9Al_4, $C_{Al} = 4/13 \approx 0.31$, $C_{Cu} = 9/13 \approx 0.69$.

We can calculate the mutual diffusion coefficient for each phase at temperature 160°C by the Darken equation [8, 10] and taking into account Eqs. (4)–(6):

$$D_i^{*} = C_{Al} D_{Cu}^{*} + C_{Cu} D_{Al}^{*}; i = 1, 2, 3; D_1^{*} = 6.64 \cdot 10^{-20} m^2/s; D_2^{*} = 1.3 \cdot 10^{-19} m^2/s; D_3^{*} = 1.8 \cdot 10^{-21} m^2/s.$$

We can calculate using the methods described in [4, 11]: $K_{123}(T_7 = 160°C) \approx 2.8 \cdot 10^{-6}$ μm²/s, $t_0(T = 160^oC) \approx \frac{2X_{Cu}^2}{K_{123}} \approx 5 years$, so the problem remains unsolved.

It was founded experimentally, that copper electrochemical corrosion is higher than aluminium electrochemical corrosion in approximately two times at room temperature [4, 11], so a thin Al layer can prevent copper electrochemical corrosion. It was reported also about the influence of hydrogen and the absence of a passive layer on the corrosive properties of aluminium alloys [12].

Besides, the soldered copper/tin-based contacts are the weakest part of the chip that can be related to intermetallics and the Kirkendall-Frenkel porosity formation in the contact zone [13]. One of the most common reasons for chip failure is the soldered. The typical range of packaging and operation of the integrated circuits is from room temperature to 250°C [14].

Hydrostatic pressure of Argon gas (≈10 MPa) can decrease Kirkendal-Frenkel porosity formation, but practically cannot decrease mutual diffusion coefficients, but hot isostatic pressing (p ≈ 100 MPa, Argon) removes porosity due to homogenisation heat treatment in alloy CMSX4 and superalloy CMSX10 [15].

It was clarified also that carbon steel-stainless steel with the environment of flowing sodium chloride does indeed produce synergetic corrosion instead of antagonistic corrosion [16].

Electric current can destruct wire bonding in microelectronics packaging, so we planned to investigate copper, iron, and aluminium electrochemical corrosion at room temperature and temperature 100°C. Direct current can dissolve metal anode into electrolyte, and we planned to do experiments under the same conditions: initial radii of Cu, Fe, and Al anodes should be approximately equal, electrolyte concentration should be the same, anodes lengths immersed into electrolyte should be equal, graphite cathodes should be the same, direct electric current value should be practically the same. Aluminium can dissolve into electrolyte only as of the Al^{3+} ions, so the charge of aluminium ions should be exactly equal to 3, but copper can dissolve into electrolyte as the Cu^{+} ions and the Cu^{2+} ions, and the charge of copper ions could be equal to 1 or 2, and iron can dissolve into electrolyte as the Fe^{2+} ions and the Fe^{3+} ions, and the charge of iron ions could be equal to 2 or 3. We need to find appropriate mathematical equations to calculate the charges of copper, iron, and aluminium ions dissolved into NaCl solution.

2. Experimental results of copper, iron, and aluminium electrochemical corrosion

2.1 Investigation at room temperature

Cylindrical anodes (99.99% Cu, 99.96% Fe, and 99.99% Al) were used for copper and aluminium [4, 11], and also iron electrochemical corrosion investigation. Sodium chloride (NaCl) solution was used as an electrolyte (**Figure 2**).

Direct electric current and anode mass decreasing were measured. First of all, we need to be assured that the Cu^{+} ions (or the Cu^{2+}), the Fe^{2+} (or the Fe^{3+}), and the Al^{3+} were present in NaCl solution. The rate of anode dissolving into electrolyte can be calculated using Faraday's law of electrolysis:

$$\frac{dm}{dt} = \frac{MI}{zF}, dm = \rho \cdot L \cdot \pi \cdot d\left(R^2(t)\right). \tag{7}$$

Here, m is anode mass dissolved into the electrolyte, t is a time of the experiment, M is molar mass, I is the direct electric current value, F is the Faraday constant ($F \approx 96{,}500$ C mol^{-1}), z is a charge of ions, R is anode radius, L is anode length immersed into the electrolyte. Electric current value did not change, so one can calculate:

$$z = \frac{MIt}{F\pi\rho L\left(R^2(t=0) - R^2(t)\right)}, \tag{8}$$

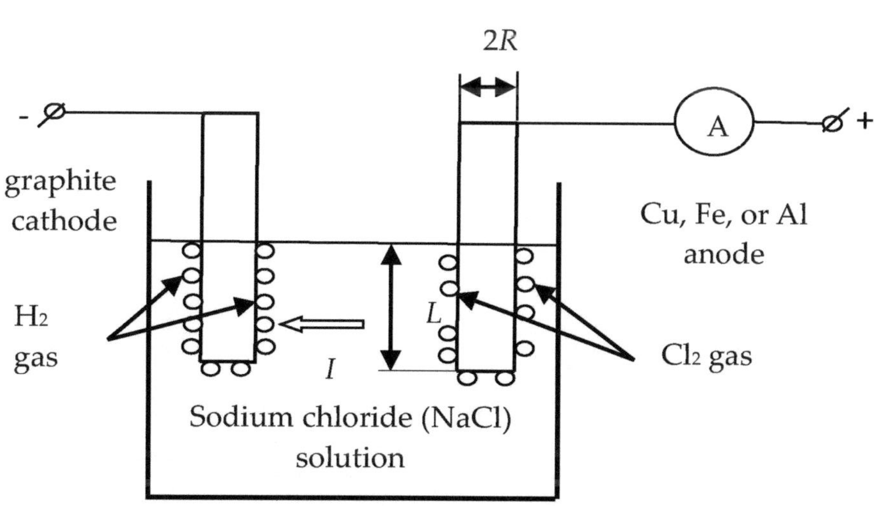

Figure 2.
Scheme of experimental equipment at room temperature. Cu, Fe, and Al anodes dissolve into NaCl solution as Cu^{+}, Fe^{2+}, and Al^{3+} ions.

where ρ is anode density. Charges of copper, iron, and aluminium ions were calculated:

$$z_{Cu} = \frac{63.55 \cdot 10^{-3} kg/_{mol} \cdot 2.8A \cdot 1.2 \cdot 10^3 s}{F \cdot \pi \cdot 8.9 \cdot 10^3 kg/_{m^3} L_{Cu} \cdot \left(R_{Cu}^2(t=0) - R_{Cu}^2(t_4)\right)} \approx 0.995 \approx 1, \tag{9}$$

$$z_{Al} = \frac{27 \cdot 10^{-3} kg/_{mol} \cdot 3.1A \cdot 1.2 \cdot 10^3 s}{F \cdot \pi \cdot 2.7 \cdot 10^3 kg/_{m^3} L_{Al} \cdot \left(R_{Al}^2(t=0) - R_{Al}^2(t_4)\right)} \approx 2.954 \approx 3, \tag{10}$$

$$z_{Fe} = \frac{55.847 \cdot 10^{-3} kg/_{mol} \cdot 3.15A \cdot 1.2 \cdot 10^3 s}{F \cdot \pi \cdot 7.86 \cdot 10^3 kg/_{m^3} L_{Fe} \cdot \left(R_{Fe}^2(t=0) - R_{Fe}^2(t_4)\right)} \approx 2.03 \approx 2, \tag{11}$$

where $L_{Cu} \approx L_{Al} \approx L_{Fe}$: $L_{Cu} = L_{Fe} = 5 \cdot 10^{-2}$ m, $L_{Al} = 4.5 \cdot 10^{-2}$ m; $R_{0Cu} = R_{0Al} = 2.8$ mm, $R_{0Fe} = 2.98$ mm; $I_{Al} \approx I_{Cu} \approx I_{Fe}$: $I_{Fe} = 3.15$A, $I_{Al} = 3.1$A, $I_{Cu} = 2.8$A, so copper dissolved into NaCl solution as the Cu^+ ions, iron dissolved into NaCl solution as the Fe^{2+} ions, and aluminium dissolved into NaCl solution as the Al^{3+} ions. Anodes radii-decreasing kinetics is shown in **Figure 3**. Experiments were carried during $t_1 = 5$ min, $t_2 = 10$ min, $t_3 = 15$ min, and $t_4 = 20$ min. Experimental results are as follows: $R_{1Cu} = 2.74$ mm, $R_{2Cu} = 2.67$ mm, $R_{3Cu} = 2.59$ mm, $R_{4Cu} = 2.5$ mm; $R_{1Al} = 2.77$ mm, $R_{2Al} = 2.73$ mm, $R_{3Al} = 2.68$ mm, $R_{4Al} = 2.62$ mm, $R_{1Fe} = 2.95$ mm, $R_{2Fe} = 2.92$ mm, $R_{3Fe} = 2.88$ mm, $R_{4Fe} = 2.83$ mm. Measurement precision was 0.01 mm or 10 micrometres.

Chemical reactions took place near the positive electrode (anode):

$$Cu^0 - e^- = Cu^+, Al^0 - 3e^- = Al^{3+}, Fe^0 - 2e^- = Fe^{2+};$$

$$Cu^+ + Cl^- = CuCl\downarrow, Al^{3+} + 3Cl^- = AlCl_3\downarrow, Fe^{2+} + 2Cl^- = FeCl_2\downarrow; \tag{12}$$

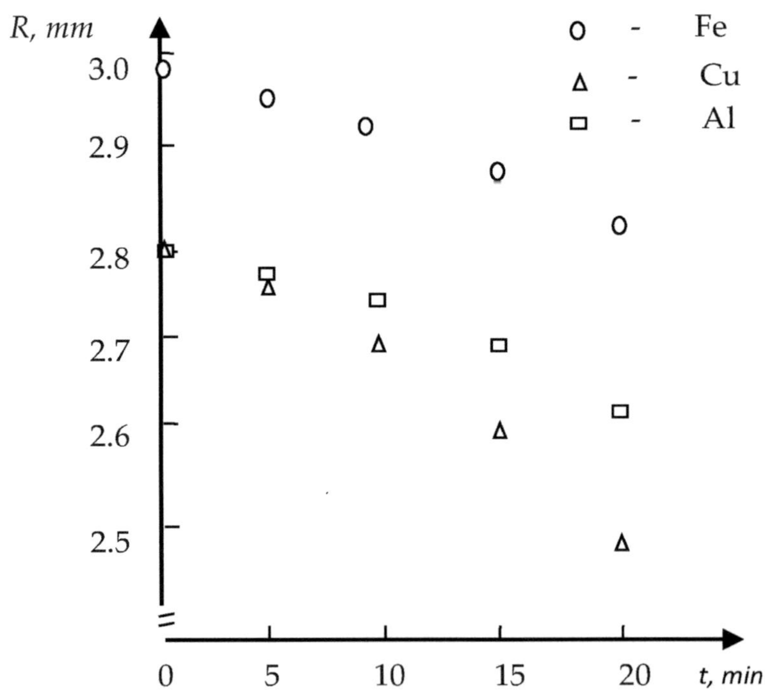

Figure 3.
Cu, Fe, and Al anodes radii decreasing kinetics at room temperature.

$$Cl^- - e^- = Cl^0, Cl^0 + Cl^0 = Cl_2 \uparrow .$$

Chlorine gas was formed near the anode.

Chemical reactions took place near the negative electrode (cathode):

$$Na^+ - e^- = Na^0, 2Na + 2H_2O = 2NaOH + H_2 \uparrow . \quad (13)$$

Hydrogen gas was formed near the cathode.

Anodes radii decreasing rate constants can be calculated as the average value of four experiments to increase calculation precise:

$$k_{Cu} = \frac{4R_0^2 - \sum_{i=1}^4 R_i^2}{\sum_{i=1}^4 t_i} \approx 1.25 \cdot 10^{-9} {}^{m^2}\!/\!_{s}, k_{Al} = \frac{4R_0^2 - \sum_{i=1}^4 R_i^2}{\sum_{i=1}^4 t_i} \approx 7.29 \cdot 10^{-10} {}^{m^2}\!/\!_{s}, \quad (14)$$

$$k_{Fe} = \frac{4R_{0Fe}^2 - \sum_{i=1}^4 R_{iFe}^2}{\sum_{i=1}^4 t_i} \approx 7.26 \cdot 10^{-10} {}^{m^2}\!/\!_{s}, k_{Cu} \approx 1.71 k_{Al}; k_{Al} \approx k_{Fe},$$

so copper electrochemical corrosion is much higher than aluminium and iron electrochemical corrosion, despite $I_{Fe} \approx I_{Al} \geq I_{Cu}$: $I_{Fe} \approx I_{Al} \approx 1.1 I_{Cu}$. It needs to point out that k_{Cu}, k_{Fe}, and k_{Al} have dimensionalities as diffusion coefficients [m^2/s], because electrochemical corrosion occurs through anodes' surface.

2.2 Investigation at temperature 100°C

Experiments were carried also at temperature 100°C. Cylindrical anodes (99.99% Cu, 99.99% Al, and 99.96% Fe) were used for copper and aluminium [17] and also iron electric corrosion investigation. Sodium chloride (NaCl) solution was used as an electrolyte (**Figure 4**). Direct electric current and anodes' mass decreasing rate were measured (**Figure 5**).

Electric current value did not change, so one can calculate the following:

$$z_{Cu} = \frac{63.55 \cdot 10^{-3} {}^{kg}\!/\!_{mol} \cdot 3.05A \cdot 1.2 \cdot 10^3 s}{F \cdot \pi \cdot 8.9 \cdot 10^3 {}^{kg}\!/\!_{m^3} L_{Cu} \cdot \left(R_{Cu}^2(t=0) - R_{Cu}^2(t_4)\right)} \approx 1.47 \approx \frac{1+2}{2}, \quad (15)$$

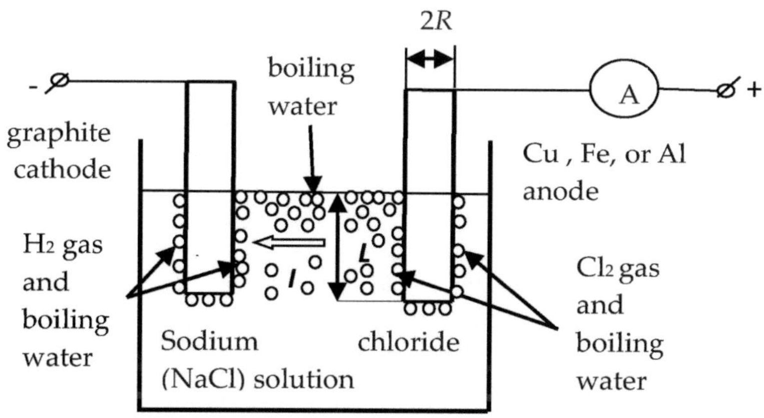

Figure 4.
Scheme of experimental equipment at T = 100°C. Cu, Fe, and Al anodes dissolved into NaCl solution as Cu^+, cu^{2+}, Fe^{2+}, and Al^{3+} ions.

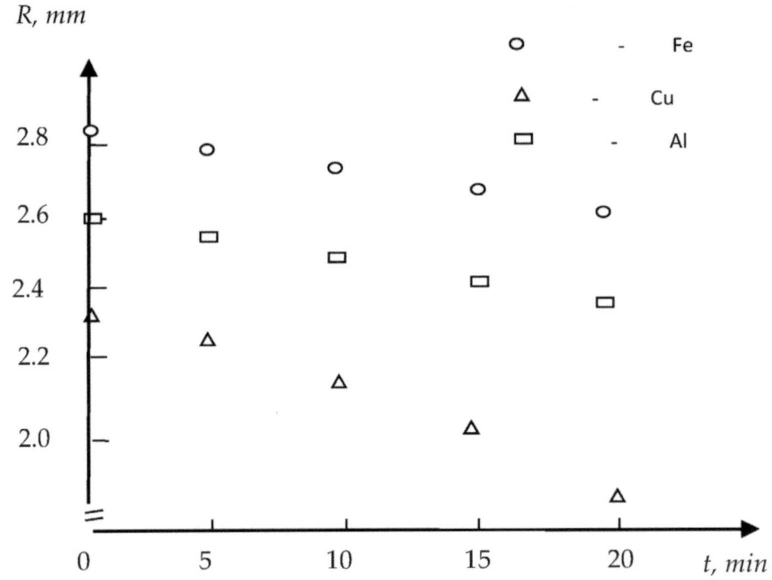

Figure 5.
Cu, Fe, and Al anodes radii decreasing kinetics at T = 100°C.

$$z_{Al} = \frac{27 \cdot 10^{-3} kg/_{mol} \cdot 3.15A \cdot 1.2 \cdot 10^3 s}{F \cdot \pi \cdot 2.7 \cdot 10^3 kg/_{m^3} L_{Al} \cdot \left(R_{Al}^2(t=0) - R_{Al}^2(t_4)\right)} \approx 2.85 \approx 3, \tag{16}$$

$$z_{Fe} = \frac{55.847 \cdot 10^{-3} kg/_{mol} \cdot 3.15A \cdot 1.2 \cdot 10^3 s}{F \cdot \pi \cdot 7.86 \cdot 10^3 kg/_{m^3} L_{Fe} \cdot \left(R_{Fe}^2(t=0) - R_{Fe}^2(t_4)\right)} \approx 2.01 \approx 2, \tag{17}$$

where $L_{Cu} = L_{Al} = 4{\cdot}10^{-2}$ m, $L_{Fe} = 5{\cdot}10^{-2}$ m, R_{0Cu} = 2.27 mm, R_{0Al} = 2.6 mm, R_{0Fe} = 2.83 mm, I_{Al} = 3.15 A, I_{Fe} = 3.13 A, I_{Cu} = 3.05 A, so copper dissolved into NaCl solution as Cu^+ and Cu^{2+} ions (copper dissolved into NaCl solution as Cu^+ ions at room temperature), iron dissolved into NaCl solution as the Fe^{2+} ions (as at room temperature), and aluminium dissolved into NaCl solution as Al^{3+} ions (as at room temperature). Anode radii-decreasing kinetics is shown in **Figure 5**. Experiments were carried during t_1 = 5 min, t_2 = 10 min, t_3 = 15 min, and t_4 = 20 min. Experimental results are as follows: R_{1Cu} = 2.2 mm, R_{2Cu} = 2.12 mm, R_{3Cu} = 2.03 mm, R_{4Cu} = 1.92 mm; R_{1Al} = 2.56 mm, R_{2Al} = 2.51 mm, R_{3Al} = 2.45 mm, R_{4Al} = 2.38 mm; R_{1Fe} = 2.80 mm, R_{2Fe} = 2.76 mm, R_{3Fe} = 2.72 mm, R_{4Fe} = 2.67 mm. Measurement precision was 0.01 mm or 10 micrometres. We carried additional experiments, but the result was the same.

Chemical reactions are more complicated at 100°C than at room temperature near positive electrodes (anodes):

$$Cu^0 - e^- = Cu^+, Cu^0 - 2e^- = Cu^{2+}, Al^0 - 3e^- = Al^{3+}, Fe^0 - 2e^- = Fe^{2+};$$
$$Cu^+ + Cl^- = CuCl\downarrow, Al^{3+} + 3Cl^- = AlCl_3\downarrow, Fe^{2+} + 2Cl^- = FeCl_2\downarrow; \tag{18}$$
$$Cu^{2+} + 2Cl^- = CuCl_2\downarrow,\ Cl^- - e^- = Cl^0, Cl^0 + Cl^0 = Cl_2\uparrow.$$

Chlorine gas and boiling water were formed near anodes.

Chemical reactions took place near negative electrodes (cathodes):

$$Na^+ - e^- = Na^0, 2Na + 2H_2O = 2NaOH + H_2\uparrow. \tag{19}$$

Hydrogen gas and boiling water were formed near cathodes.

Anode radii-decreasing rate constants can be calculated as average value of four experiments to increase calculation precise:

$$k_{Cu} = \frac{4R_{0Cu}^2 - \sum_{i=1}^{4} R_{iCu}^2}{\sum_{i=1}^{4} t_i} \approx 1.154 \cdot 10^{-9} m^2/s, \{1.25.10^{-9} \text{ at room temperature}\},$$

$$k_{Al} = \frac{4R_{0Al}^2 - \sum_{i=1}^{4} R_{iAl}^2}{\sum_{i=1}^{4} t_i} \approx 8.42 \cdot 10^{-10} m^2/s, \{7.29.10^{-10} \text{ at room temperature}\},$$

(20)

$$k_{Fe} = \frac{4R_{0Fe}^2 - \sum_{i=1}^{4} R_{iFe}^2}{\sum_{i=1}^{4} t_i} \approx 6.83 \cdot 10^{-10} m^2/s, \{7.23.10^{-10} \text{ at room temperature}\},$$

$$k_{Cu} \approx 1.37 k_{Al}, \{1.72 \text{ at room temperature } T_1 \approx 27°C\},$$

so copper electrochemical corrosion is higher at room temperature $T_1 \approx 27°C$, aluminium electrochemical corrosion is higher at temperature T_2 = 100°C, and the ratio of electrochemical corrosion rates, k_{Cu}/k_{Al}, decreases with temperature increasing, although iron electrochemical corrosion rate practically does not depend on temperature below 100°C. It is obvious, because of the higher melting point of iron than the melting point of copper or aluminium. We can conclude that the Cu^{2+} ions are less mobile than Cu^+ ions. It needs to point out that k_{Cu}, k_{Al}, and k_{Fe} have dimensionalities as diffusion coefficients, D^*_{Cu}, D^*_{Al}, D^*_{Fe} [m²/s], because electrochemical corrosion occurs through anodes' surface.

Dislocation pipe and volume diffusion activation energies can be calculated in such a way. The Arrhenius law is valid for dislocation pipe diffusion and volume diffusion in ultra-high-purity samples [8, 18]:

$$D_d^* = D_{0d} e^{-Q_d/(RT)} \text{ or } D_d^* = D_{0d} e^{-E_d/(k_B T)}, \text{ and } D_V^* = D_{0V} e^{-Q_V/(RT)} \text{ or } D_V^* = D_{0V} e^{-E_V/(k_B T)},$$

(21)

$$Q[J/mol] = F \cdot E[eV]. \tag{22}$$

Here, $R \approx 8.314$ JK^{-1} is the gas constant, k_B is the Boltzmann constant, $Q_d(E_d)$ is the dislocation pipe diffusion activation energy ($Q_d = FE_d$), $F \approx 96{,}500$ Cmol^{-1} is the Faraday constant, $Q_V(E_V)$ is the volume diffusion activation energy ($Q_V = FE_V$), D_{0d} and D_{0V} are the pre-exponential factors, T is the absolute temperature.

Our experimental results allow us to calculate:

$$\frac{k_{Cu}}{k_{Al}}(T) \approx \frac{D_{Cu}^*}{D_{Al}^*}(T) = \frac{D_{0Cu}^*}{D_{0Al}^*} e^{(Q_{Al} - Q_{Cu})/(RT)}; \ln\left(\frac{D_{0Cu}^*}{D_{0Al}^*}\right) = -0.6; Q_{Al} - Q_{Cu} = 2.9 kJ/mol,$$

(23)

$$\frac{D_{0Cu}^*}{D_{0Al}^*} = 0.55, \frac{D_{Cu}^*}{D_{Al}^*}(T_3) = 1 \Rightarrow T_3 = \frac{2900 J/mol}{0.6R} \approx 583K \approx 310^o C, \tag{24}$$

so diffusion activation energy of Al, Q_{Al}, is higher than the diffusion activation energy of Cu, Q_{Cu}, ($Q_{Al} > Q_{Cu}$, Q_{Al}-Q_{Cu} = 2.9 kJ/mol), at temperatures from 20 to 100°C, because the Cu^+ ions have higher mobilities than the Al^{3+} ions, and copper electrochemical corrosion rate can be approximately equal to aluminium electrochemical corrosion at temperature about $T_3 \approx 300°C$ due to the Cu^{2+} ions are less mobile than the Cu^+ ions. Moreover, the pre-exponential factors are approximately the same: $D^*_{0Al} \approx 2D^*_{0Cu}$.

3. Intrinsic diffusivities ratio and diffusion activation energy calculations

3.1 Intrinsic diffusivities ratio of Cu and Al analysis

We can analyse described the experimental results in the Al-Cu system for bulk samples [19] since the ratio D^*_{Al}/D^*_{Cu} was not calculated in [19]:

$$\frac{D^*_{Cu}}{D^*_{Al}} \approx \frac{\sum_{j=1}^{N} X_j - X_K(1-C_i)\sqrt{\pi}}{\sum\limits_{j=1}^{N} X_j + C_i X_K \sqrt{\pi}} < 1, C_i = C_{Al}, \tag{25}$$

where N is formed phases quantity, X_j is phase j's thickness, C_i is the average concentration of aluminium in phase i, and X_K is the Kirkendall shift length.

Five phases are formed in the Al-Cu system at temperatures from 400 to 535°C in bulk samples [19]: θ-phase (phase 1) $CuAl_2$ (C_1 = 2/3), η_2-phase (phase 2) CuAl (C_2 = 1/2), ζ_2-phase (phase 3b) Cu_4Al_3 (C_{3b} = 3/7), δ-phase (phase 3a) Cu_3Al_2 (C_{3a} = 2/5), and γ_2-phase (phase 3) Cu_9Al_4 ($C_3 = 4/13 \approx 0.31 \approx 1/3$, $C = C_{Al}$). Inert markers were in δ-phase (phase 3a) Cu_3Al_2 (C_{3a} = 2/5 = 0.4) and moved to Al side during mutual diffusion. In general, inert markers move to the faster diffusivity component side. We can calculate (C_{3a} = 0.4):

$$\frac{D^*_{Cu}}{D^*_{Al}}(T_1 = 535^oC) \approx \frac{X_1 + X_2 + X_{3b} + X_{3a} + X_3 - X_K 0.6\sqrt{\pi}}{X_1 + X_2 + X_{3b} + X_{3a} + X_3 + 0.4 X_K\sqrt{\pi}} \approx 0.814;$$
$$y_1 = \frac{D^*_{Al}}{D^*_{Cu}} = \frac{1}{0.814} \approx 1.228, \tag{26}$$

T_1 = 535°C = 808 K, t = 40 h, $X_K \approx 20.5$ μm, $X_1 + X_2 + X_{3b} + X_{3a} + X_3 \approx 180$ μm;

$$\frac{D^*_{Cu}}{D^*_{Al}}(T_2 = 515^oC) \approx \frac{X_1 + X_2 + X_{3b} + X_{3a} + X_3 - X_K(1-C_{3a})\sqrt{\pi}}{X_1 + X_2 + X_{3b} + X_{3a} + X_3 + C_{3a} X_K\sqrt{\pi}} \approx 0.856;$$
$$y_2 = \frac{D^*_{Al}}{D^*_{Cu}} \approx 1.168, \tag{27}$$

T_2 = 515°C = 788 K, t = 40 h, $X_K \approx 11$ μm, $X_1 + X_2 + X_{3b} + X_{3a} + X_3 \approx 127$ μm);

$$\frac{D^*_{Cu}}{D^*_{Al}}(T_3 = 495^oC) \approx \frac{X_1 + X_2 + X_{3b} + X_{3a} + X_3 - X_K 0.6\sqrt{\pi}}{X_1 + X_2 + X_{3b} + X_{3a} + X_3 + 0.4 X_K\sqrt{\pi}} \approx 0.916;$$
$$y_3 = \frac{1}{0.916} \approx 1.092, \tag{28}$$

T_3 = 495°C = 768 K, t = 40 h, $X_K \approx 5$ μm, $X_1 + X_2 + X_{3b} + X_{3a} + X_3 \approx 101$ μm;

$$\frac{D^*_{Cu}}{D^*_{Al}}(T_4 = 475^oC) \approx \frac{X_1 + X_2 + X_{3b} + X_{3a} + X_3 - X_K(1-0.4)\sqrt{\pi}}{X_1 + X_2 + X_{3b} + X_{3a} + X_3 + 0.4C X_K\sqrt{\pi}} \approx 0.969;$$
$$y_4 = \frac{D^*_{Al}}{D^*_{Cu}} \approx 1.032, \tag{29}$$

T_4 = 475°C = 748 K, t = 90 h, $X_K \approx 2$ μm, $X_1 + X_2 + X_{3b} + X_{3a} + X_3 \approx 113$ μm.

We can use these four points to calculate by the least square method to increase calculation precise:

$$\Delta Q = Q_{Al} - Q_{Cu} = \frac{4\sum_{j=1}^{4}\left(\frac{1000}{RT_j}\ln y_j\right) - \sum_{j=1}^{4}\ln y_j \sum_{j=1}^{4}\frac{1000}{RT_j}}{4\sum_{j=1}^{4}\left(\frac{1000}{RT_j}\right)^2 - \left(\sum_{j=1}^{4}\frac{1000}{RT_j}\right)^2} \approx -13.4kJ/mol, \tag{30}$$

$$y_0 = \exp\frac{\sum_{j=1}^{4}\left(\frac{1000}{RT_j}\right)^2\sum_{j=1}^{4}\ln y_j - \sum_{j=1}^{4}\frac{1000}{RT_j}\sum_{j=1}^{4}\left(\frac{1000}{RT_j}\ln y_j\right)}{4\sum_{j=1}^{4}\left(\frac{1000}{RT_j}\right)^2 - \left(\sum_{j=1}^{4}\frac{1000}{RT_j}\right)^2} \approx \exp(2.2) \approx 9, \tag{31}$$

$$\frac{D^*_{Al}}{D^*_{Cu}}(T) = \frac{D^*_{0Al}}{D^*_{0Cu}} e^{(Q_{Al}-Q_{Cu})/(RT)} \approx 9e^{-13.4kJmol^{-1}/(RT)}, \tag{32}$$

$$\frac{D^*_{Al}}{D^*_{Cu}}(T_5) = y_5 = 1 \Rightarrow T_5 = \frac{13400J/mol}{R\ln 9} \approx 733K \approx 460^oC, \tag{33}$$

so $Q_{Al} < Q_{Cu}$ (Q_{Al}-Q_{Cu} = −13.4 kJ/mol) because the Cu^{2+} ions have less mobilities than the Al^{3+} ions, and we can conclude that the Kirkendall displacement changes sign at a temperature about $T_5 \approx 460°C$ for bulk samples. The pre-exponential factors are different in nine times: $D^*_{0Al} \approx 9D^*_{0Cu}$.

Diffusion activation energy of Al is less than the diffusion activation energy of Cu ($Q_{Al} < Q_{Cu}$) at temperatures from 160–250°C for mutual diffusion in copper-a luminium thin film double layers, but the pre-exponential factors are different in 10 times [9]. Isolated W islands, 150 Å in diameter, have been deposited between Cu and Al thin film double layers to serve as inert diffusion markers. Marker displacements have been measured. We can calculate the ratio D^*_{Al}/D^*_{Cu} for each phase at different temperatures:

$$\frac{D^*_{Cu}}{D^*_{Al}}(T) = \frac{D^*_{0Cu}}{D^*_{0Al}} e^{(Q_{Al}-Q_{Cu})/(RT)} = 24e^{-14kJmol^{-1}/(RT)} \text{ in } \theta\text{-phase (phase 1) } CuAl_2, C_{Al} = 2/3 \approx 0.67,$$

$$\frac{D^*_{1Cu}}{D^*_{1Al}}(T = 250^oC = 523K) = 24e^{-14000Jmol^{-1}/(8.314x523)} \approx 24e^{-3.22} \approx 0.96,$$

$$\frac{D^*_{1Cu}}{D^*_{1Al}}(T = 160^oC = 433K) \approx 24e^{-3.89} \approx 0.49,$$

so the Al atoms diffuse faster than the Cu atoms in θ-phase at temperatures from 160 to 250°C;

$$\frac{D^*_{Cu}}{D^*_{Al}}(T) = \frac{D^*_{0Cu}}{D^*_{0Al}} e^{(Q_{Al}-Q_{Cu})/(RT)} = 7\cdot 10^4 e^{-38kJmol^{-1}/(RT)} \text{ in } \eta_2\text{-phase (phase 2) CuAl}, C_{Al} = 1/2 = 0.5,$$

$$\frac{D^*_{2Cu}}{D^*_{2Al}}(T = 250^oC) = 7\cdot 10^4 e^{-38000Jmol^{-1}/(RT)} \approx 11.2, \frac{D^*_{2Cu}}{D^*_{2Al}}(T = 160^oC) \approx 1.8,$$

so the Cu atoms diffuse faster than the Al atoms in η2-phase at temperatures from 160 to 250°C;

$$\frac{D^*_{Cu}}{D^*_{Al}}(T) = \frac{D^*_{0Cu}}{D^*_{0Al}} e^{(Q_{Al}-Q_{Cu})/(RT)} = 14e^{-9kJmol^{-1}/(RT)} \text{ in } \gamma_2\text{-phase (phase 3) } Cu_9Al_4, C_{Al} = 4/13 \approx 0.31,$$

$$\frac{D^*_{3Cu}}{D^*_{3Al}}(T = 250^oC) = 14e^{-9000Jmol^{-1}/(RT)} \approx 1.77, \frac{D^*_{3Cu}}{D^*_{3Al}}(T = 160^oC) = 14e^{-9kJmol^{-1}/(RT)} \approx 1.15,$$

so the Cu atoms diffuse faster than the Al atoms in γ_2-phase at temperatures from 160 to 250°C.

The Cu-rich phases can be formed faster than the Al-rich phases at temperatures from 160 to 250°C, and the Cu atoms can diffuse faster than the Al atoms in the Al-Cu system at temperatures from 160 to 250°C. The Al-rich phases can be formed faster than the Cu-rich phases at temperatures from 400 to 535°C, and the Al atoms can diffuse faster than the Cu atoms in the Al-Cu system at temperatures from 400 to 535°C. It depends on the crystal structure of each phase, but, in general, it could depends on conclusions that the Cu^{2+} ions are less mobile than the Cu^+ ions, and the ratio D^*Al/D^*Cu depends on temperature.

3.2 Diffusion activation energy calculation

3.2.1 Diffusion activation energy calculation in the Cu-Al system

Mutual diffusion coefficients were calculated for all five phases [19]:

$$\tilde{D}^*_1 = 5.6 \cdot 10^{-5} e^{-127.6kJmol^{-1}/(RT)} m^2/s; \tilde{D}^*_2 = 2.2 \cdot 10^{-4} e^{-148.5kJmol^{-1}/(RT)} m^2/s; \quad (34)$$

$$\tilde{D}^*_{3b} = 1.6 \cdot 10^2 e^{-230.5kJmol^{-1}/(RT)} m^2/s, \tilde{D}^*_{3a} = 2.1 \cdot 10^{-4} e^{-138.1kJmol^{-1}/(RT)} m^2/s, \tilde{D}^*_3 = 8.5 \cdot 10^{-5} e^{-136kJmol^{-1}/(RT)} m^2/s.$$

We can see that $Q_1 < Q_2$, $Q_1 < Q_3$, and $Q_3 < Q_2$ because of $K_1 > K_2$, $K_1 > K_3$, and $K_3 > K_2$, and $D_{01} \approx D_{02} \approx D_{03}$. Phase j's rate formation is K_j. Three phases are formed in the Al-Cu system at temperatures 300 and 350°C [5]: $CuAl_2$, CuAl, and Cu_9Al_4. Phases formation rates were experimentally measured: K_1 = 860x10^{-18} m^2/s, K_2 = 100x10^{-18} m^2/s, and K_3 = 360x10^{-18} m^2/s at temperature 350°C; K_1 = 77x10^{-18} m^2/s, K_2 = 18x10^{-18} m^2/s, and K_3 = 35x10^{-18} m^2/s at temperature 300°C, so $K_1 > K_2$, $K_1 > K_3$, and $K_3 > K_2$. We can calculate assuming C_1 = 2/3, C_2 = 1/2, C_3 = 1/3, $C = C_{Al}$ [11]:

$$D_1 \approx \frac{1}{2}\left(C_1(1-C_1)K_1 + C_2(1-C_1)\sqrt{K_1K_2} + C_3(1-C_1)\sqrt{K_1K_3}\right); \quad (35)$$

$$D_2 \approx \frac{1}{2}\left(C_2(1-C_2)K_2 + C_2(1-C_1)\sqrt{K_1K_2} + C_3(1-C_2)\sqrt{K_2K_3}\right); \quad (36)$$

$$D_3 \approx \frac{1}{2}\left(C_3(1-C_3)K_3 + C_3(1-C_1)\sqrt{K_1K_3} + C_3(1-C_2)\sqrt{K_2K_3}\right); \quad (37)$$

$$D_1(T_2 = 350^0C) \approx \frac{1}{9}K_1 + \frac{1}{12}\sqrt{K_1K_2} + \frac{1}{18}\sqrt{K_1K_3} \approx 150 \cdot 10^{-18} m^2/s, D_1(T_1 = 300^oC) \approx 15 \cdot 10^{-18} m^2/s; \quad (38)$$

$$D_2(T_2 = 350^oC) \approx \frac{1}{8}K_2 + \frac{1}{12}\sqrt{K_1K_2} + \frac{1}{12}\sqrt{K_2K_3} \approx 50x10^{-18}m^2/s, D_2(T_1 = 300^oC) \approx 8 \cdot 10^{-18}m^2/s; \quad (39)$$

$$D_3(T_2 = 350^oC) \approx \frac{1}{9}K_3 + \frac{1}{18}\sqrt{K_1K_3} + \frac{1}{12}\sqrt{K_2K_3} \approx 90 \cdot 10^{-18}m^2/s, D_3(T_1 = 300^oC) \approx 12 \cdot 10^{-18}m^2/s. \quad (40)$$

Moisy et al. [5] did not calculate diffusion activation energies and the pre-exponential factors, so we can do it:

$$Q_i = \frac{RT_1T_2}{T_2 - T_1} \ln\left(\frac{D_i(T_2)}{D_i(T_1)}\right), D_{0i} = D_i(T_1)e^{Q_i/(RT_1)} = D_i(T_2)e^{Q_i/(RT_2)}; \quad (41)$$

$$\tilde{D}_1 = 4.3 \cdot 10^{-5}e^{-136.7kJmol^{-1}/(RT)}m^2/s, \tilde{D}_2 = 6.6 \cdot 10^{-8}e^{-108.8kJmol^{-1}/(RT)}m^2/s, \quad (42)$$
$$\tilde{D}_3 = 9.6 \cdot 10^{-7}e^{-119.6kJmol^{-1}/(RT)}m^2/s.$$

Eq. (42) correspond to Eq. (34). We can use several, N, points to calculate by the least square method to increase calculation precise:

$$Q_i = -\frac{N\sum_{j=1}^{N}\left(\frac{1000}{RT_j}\ln D_i(T_j)\right) - \sum_{j=1}^{N}\ln D_i(T_j)\sum_{j=1}^{N}\frac{1000}{RT_j}}{N\sum_{j=1}^{N}\left(\frac{1000}{RT_j}\right)^2 - \left(\sum_{j=1}^{N}\frac{1000}{RT_j}\right)^2}[kJ/mol], \quad (43)$$

$$D_{0i} = \exp\frac{\sum_{j=1}^{5}\left(\frac{1000}{RT_j}\right)^2\sum_{j=1}^{5}\ln D_i(T_j) - \sum_{j=1}^{5}\frac{1000}{RT_j}\sum_{j=1}^{4}\left(\frac{1000}{RT_j}\ln D_i(T_j)\right)}{5\sum_{j=1}^{5}\left(\frac{1000}{RT_j}\right)^2 - \left(\sum_{j=1}^{5}\frac{1000}{RT_j}\right)^2}\left[m^2/s\right]. \quad (44)$$

Eqs. (43) and (44) give Eq. (41) for only two points (N = 2).

3.2.2 Diffusion activation energy calculation in pure iron

A method of dislocation pipe diffusion parameter determination during the type B diffusion kinetics was suggested by the model of dislocation pipe diffusion involving outflow [6, 20]. The method involves diffusion dislocation pipe kinetics for two different annealing times at the same temperature during the type B kinetics and dislocation pipe kinetics for one annealing time at other lower temperature during the type C kinetics. Transition time for type B kinetics to type A kinetics (volume diffusion) and kinetics law $t^{1/6}$ [7] for cone top rate are used in this method.

Bulk diffusion coefficients, D_V, for the diffusion of ^{59}Fe in the high-purity iron were calculated in [21] using type B → A kinetics: D_V = 1.5⋅10^{-18} m^2s^{-1} at T_1 = 973 K for $t_{B \to A}$ = 67.5ks (T_m/T_1 = 1.86, T_m is melting point of iron). Only one experiment was carried out at the same temperature for two annealing times t_1 and t_2 ($t_1 << t_2$, t_2 = 40 t_1). Dislocation diffusion coefficients for the diffusion of ^{59}Fe in the iron were calculated in [21] using type C kinetics: D_d = 3⋅10^{-16} m^2s^{-1} at T_2 = 753 K for t_C = 2.4ks (T_m/T_2 = 2.4). One can find ratio D_d/D_V: $y(t_{C \to B}) = \sqrt{\frac{D_d}{6D}}\delta$, where δ = 1 nm, $\frac{D_d}{D} = 4.3x10^6$. Ratio D_d/D_V increases remarkably for lower temperature. Dislocation pipe and volume diffusion activation energies and pre-exponential factors were not calculated in [21]. It is possible to calculate E_d and D_0: $E_d = \ln\left(\frac{D_d(T_1)}{D_d(T_2)}\right) k_B \frac{T_1 T_2}{T_1 - T_2}$, $D_0 = D_d(T_1)\exp\left(\frac{E_d}{k_B T_1}\right)$, $E_d = 1.1eV$; $Q_d = 106kJ/mol$, $D_0 = 6.85 \cdot 10^{-9} m^2 s^{-1}$. One can calculate dislocation pipe diffusion coefficient for temperature 973 K directly (T_1 = 753 K and T_2 = 693 K (type C kinetics)): $D_d \approx 10^{-14}$ m^2s^{-1}. Such value corresponds to value calculated using the proposed method. The Fisher law ($t^{1/4}$) gives $D_d \approx 10^{-16} \div 10^{-15}$ m^2s^{-1}. Such value is in two orders lower than experimentally obtained in [21]. The volume diffusion activation energy E_V can be calculated: $E_V = \ln\left(\frac{D_0}{D_V(T_1)}\right) k_B T_1$, $E_V = 1.85eV$; $Q_V \approx 179kJ/mol$. Ratio $\frac{E_d}{E_V} = 0.6$ as described in [8].

4. Conclusions

The Al atoms diffuse faster than the Cu atoms at a temperature higher than 475°C, but the Cu atoms diffuse faster than the Al atoms at a temperature lower than 100°C. The diffusion activation energy of Al is less than the diffusion activation energy of Cu at a temperature higher than 475°C, but diffusion activation energy of Cu is less than the diffusion activation energy of Al at a temperature lower than 100°C. Our investigations show that it is possible because the Cu^{2+} ions are less mobile than Cu^{+} ions.

Volume diffusion activation energy of Fe is higher than volume diffusion activation energy of Cu or Al, but dislocation pipe diffusion activation energy of Fe is smaller than volume diffusion activation energy of Cu or Al, so the Fe atoms diffuse faster along the dislocation line, but the Cu or Al atoms diffuse faster in volume.

Author details

Mykhaylo Viktorovych Yarmolenko
Faculty of Market, Information and Innovation Technologies, Kyiv National University of Technologies and Design, Cherkasy, Ukraine

*Address all correspondence to: yarmolenko.mv@knutd.edu.ua

References

[1] Kizaki T, Minho O, Kajihara M. Rate-controlling process of compound growth in Cu-Clad Al wire during isothermal annealing at 483–543 K. Materials Transactions. 2020;**61**(1): 188-194. DOI: 10.2320/matertrans.MT-M2019207

[2] Braunovic M, Alexandrov N. Intermetallic compounds at aluminum-to-copper electrical interfaces: Effect of temperature and electric current. IEEE Transactions on Components, Packaging, and Manufacturing Technology: Part A. 1994;**17**(1):78-85. DOI: 10.1109/95.296372

[3] Goh CS, Chong WLE, Lee TK, Breach C. Corrosion study and intermetallics formation in gold and copper wire bonding in microelectronics packaging. Crystals. 2013;**3**(3):391-404. DOI: 10.3390/cryst3030391

[4] Yarmolenko MV. Copper and aluminum electric corrosion investigation and intermetallics disappearance in Cu-Al system analysis. Phys. Chem. Solid St. 2020;**21**(2): 294-299. https://journals.pnu.edu.ua/index.php/pcss/article/view/3055

[5] Moisy F, Sauvage X, Hug E. Investigation of the early stage of reactive interdiffusion in the Cu-Al system by *in-situ* transmission electron microscopy. Materialia. 2020;**9**:100633. DOI: 10.1016/j.mtla.2020.100633

[6] Yarmolenko MV. Method of dislocation and bulk diffusion parameters determination. Metallofizika i Noveishie Tekhnologii. 2020;**42**(11): 1537-1546. https://mfint.imp.kiev.ua/article/v42/i11/MFiNT.42.1537.pdf

[7] Yarmolenko MV. Intermediate phase cone growth kinetics along dislocation pipes inside polycrystal grains. AIP Advances. 2018;**8**:095202. DOI: 10.1063/1.5041728

[8] Mehrer H. Diffusion in Solids. New York: Springer; 2007. 651p. http://users.encs.concordia.ca/~tmg/images/7/79/Diffusion_in_solids_Helmut_Mehrer.pdf

[9] Hentzell HTG, Tu KN. Interdiffusion in copper–aluminum thin film bilayers. II. Analysis of marker motion during sequential compound formation. Journal of Applied Physics. 1983;**54**:6929-6937. DOI: 10.1063/1.332000

[10] Darken LS. Diffusion, mobility and their interrelation through free energy in binary metallic systems. Transactions AIME. 1948;**175**:184-201. http://garfield.library.upenn.edu/classics1979/A1979HJ27500001.pdf

[11] Yarmolenko MV. Intermetallics disappearance rate analysis in double multiphase systems. DDF. 2021;**407**: 68-86. DOI: 10.4028/www.scientific.net/ddf.407.68

[12] Włodarczyk PP, Włodarczyk B. Effect of hydrogen and absence of passive layer on corrosive properties of aluminium alloys. Materials. 2020;**13**(7): 1580-1593. DOI: 10.3390/ma13071580

[13] Kumar S, Handwerker CA, Dayananda MA. Intrinsic and interdiffusion in Cu-Sn system. JPEDAV. 2011;**32**:309-319. DOI: 10.1007/s11669-011-9907-91547-7037

[14] Tu KN. Electronic Thin-Film Reliability. 1st ed. New York: Cambridge University Press; 2010. 392p. DOI:https://www.amazon.com/Electronic-Thin-Film-Reliability-King-Ning-Tu-ebook-dp-B00QIT3LXA/dp/B00QIT3LXA/ref=mt_other?_encoding=UTF8&me=&qid=

[15] Epishin A, Chyrkin A, Camin B, Saillard R, Gouy S, Viguier B. Interdiffusion in CMSX-4 related Ni-base alloy system at a supersolvus

temperature. DDF. 2021;**407**:1-10. DOI: 10.4028/www.scientific.net/ddf.407.1

[16] Prawoto Y. Synergy of erosion and galvanic effects of dissimilar steel welding: Field failure analysis case study and laboratory test results. Journal of King Saud University – Engineering Sciences. 2013;**25**:59-64. DOI: 10.1016/j.jksues.2011.12.001

[17] Yarmolenko MV. Intrinsic diffusivities ratio analysis in the Al-Cu system. Phys. Chem. Solid St. 2020; **21**(4):720-726. https://journals.pnu.edu.ua/index.php/pcss/article/view/4440

[18] Yarmolenko MV. Intermetallics disappearance rates and intrinsic diffusivities ratios analysis in the Cu-Zn and the Cu-Sn systems. Phys. Chem. Solid St. 2021;**22**(1):80-87. https://journals.pnu.edu.ua/index.php/pcss/article/view/4744

[19] Funamizu Y, Watanabe K. Interdiffusion in the Al–Cu system. Transactions of the Japan Institute of Metals. 1971;**12**(3):147-152. DOI: 10.2320/matertrans1960.12.147

[20] Yarmolenko MV. Analytically solvable differential diffusion equations describing the intermediate phase growth. Metallofizika i Noveishie Tekhnologii. 2018;**40**(9):1201-1207. https://mfint.imp.kiev.ua/article/v40/i09/MFiNT.40.1201.pdf

[21] Shima Y, Ishikawa Y, Nitta H, Yamazaki Y, Mimura K, Isshiki M, et al. Self-diffusion along dislocations in ultra high purity iron. Materials Transactions. 2002;**43**(2):173-177. https://www.jim.or.jp/journal/e/43/02/173.html

Chapter 4

Applications of the Effectiveness of Corrosion Inhibitors with Computational Methods and Molecular Dynamics Simulation

Şaban Erdoğan and Burak Tüzün

Abstract

Many experts working in the field of corrosion work in laboratories experimentally with long-term procedures and high costs by making changes in the structures of new corrosion inhibitors or existing inhibitors. Advances in computational chemistry and computer software in recent years combine corrosion prevention studies with theoretical chemistry, enabling fast, cheap and highly accurate research. Researchers working in this field can now predict the electronic, molecular and adsorption properties of anti-corrosion molecules at the molecular level with density functional theory (DFT) and Molecular Dynamics Simulation. This section includes: introduction, corrosion mechanisms, introduction to corrosion inhibitors, density functional theory (DFT) and corrosion applications, Molecular Dynamics Simulation, DFT and Molecular Dynamics Simulation applications of the effectiveness of the selected corrosion inhibitor and results. The theoretical data obtained by both the DFT approach and the molecular dynamics simulation approach showed that the corrosion inhibition efficiency order against iron corrosion for the studied Schiff bases and derivatives can be presented as: DBAMTT> SAMTT> AMTT. HOMO energy value of DBAMTT has −8,18144, HOMO energy value of SAMTT has −8,09001, and AMTT has −8,01518 in HF/6–31++G** basis set.

Keywords: Corrosion, DFT, Molecular Dynamics Simulation, Fe(110), Corrosion inhibitor

1. Introduction

Corrosion prevention studies have been intensified in recent years by the use of many organic compound classes as corrosion inhibitors for metals in acidic environments [1–3]. Both experimental and theoretical studies are carried out on this subject, but due to the fact that experimental studies are expensive and time-consuming, emphasis is placed on theoretical chemistry with software systems that have developed considerably in recent years [4, 5]. Some quantum chemical methods and molecular modeling techniques are carried out to characterize the molecular structure of the inhibitors by determining the effectiveness of corrosion inhibitors and to suggest the mechanisms of their interaction with surfaces [6–8].

Corrosion inhibitors, which are one of the easiest methods of protecting metals against corrosion, are gaining importance day by day [9–11]. The adsorption of

IntechOpen

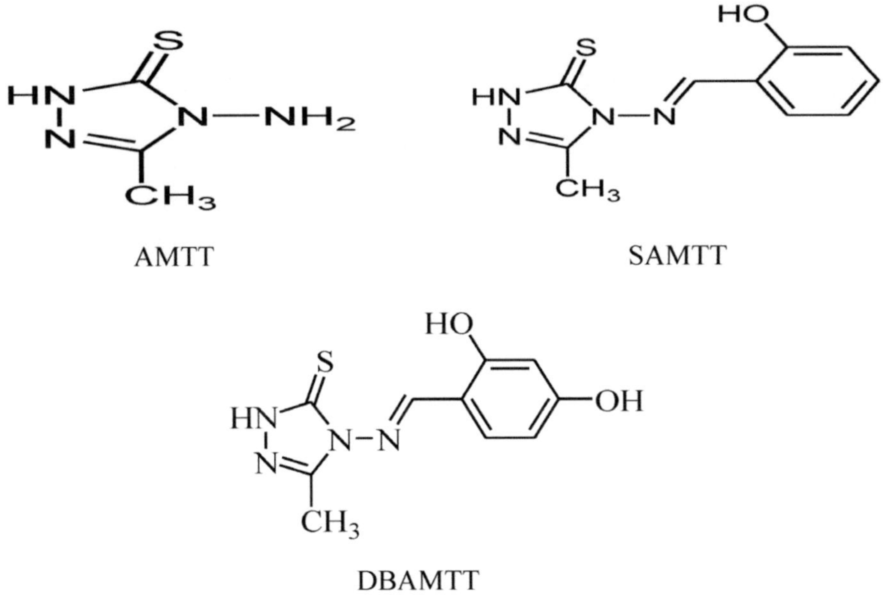

Figure 1.
Chemical molecular structures of studied Schiff bases derivatives.

these molecules depends on many physicochemical properties of the molecules [12–14]. There are many physicochemical properties for the studied molecules to be good inhibitors, such as aromaticity, steric factors, electron density etc. [15–17].

It has been stated in many studies that organic inhibitors contain heteroatoms such as nitrogen, sulfur or oxygen and that congenital double bonded heterocyclic aromatic ring systems are quite good inhibitors for mild steel [18, 19]. Schiff bases are also a very good inhibitor because they have these properties, and many experimental studies are carried out on Schiff bases. In this study, salicylideneamino-3-methyl-1,2,4-triazole-5-thi one26 (SAMTT) and 4- (2,4-dihydroxybenzylideneamino) -3-methyl-1,2,4-triazole-5-thione (DBAMTT) and methyl-1,2,4-triazole-5-thion27 (AMTT) compounds are theoretically studied [20].

The use of Conceptual density functional theory (DFT) to describe the structure and effectiveness of inhibitors in corrosion processes is becoming a well-known use. With this theory, using the energy of the highest filled molecular orbitals (E_{HOMO}) and the lowest empty molecular orbitals (E_{LUMO}), global chemical descriptors such as hardness [21], electronegativity [22], softness [23], electrophilicity [24] and chemical potential are calculated for corrosion. It provides information about the effectiveness of inhibitors. In this section, determination of the corrosion inhibition efficiency and best inhibitor of the molecules in **Figure 1** on iron corrosion is explained using quantum chemical calculations and molecular dynamics simulations approach.

2. Theory and computational details

Density Functional Theory (DFT), the most common method used to determine the chemical reactivity of molecules, aggregates, and solids, seems to be getting more popular day by day [25]. DFT calculations in this study were made with Gaussian View 5.0.8 program [26] for the preparation of Gaussian 09 [27] input files. The structures of the compounds in the study are calculated with functional B3LYP

[28, 29] based on density functional theory (DFT). High-level 6–311 ++ G (d, p) foundation sets were used in the calculations. This basic sebum is one of the most accurate basic sets. Calculations in both gas and aqueous phases were also made using SDD, 6–31 ++ G (d, p) and 6-31G base sets, as well as HF and DFT/B3LYP methods, using other levels of theory. One of the reasons to investigate the liquid phase in the study is that the corrosion is higher than the liquid phase. For the liquid phase calculations in the study, Tomasi's polarized continuity model (PCM) and self-consistent reaction area (SCRF) theory were used. These methods model the solvent as a uniform dielectric constant (DC = 78.5) continuity and define the cavity in which the solute is placed as a uniform series of interlocking atomic spheres.

In recent years, DFT methods have been found to be successful in providing insight into chemical reactivity indices such as chemical hardness (η), energy gap (ΔE_{gap} = HOMO–LUMO), electronegativity (χ), chemical potential (μ). Fukui functions *f(r)* [30] in terms of proton affinity (PA), electrophilicity (ω) and nucleophilicity (ε) and selectivity [25], spherical descriptors and local descriptors.

Reactivity indices such as electronegativity (χ), chemical hardness (η), and chemical potential (μ) are defined as derivatives of electronic energy (E) with respect to the number of electrons (N) at external potential, $\upsilon_{(r)}$. Mathematical operations related to these concepts are given through the following Equations [31].

$$\mu = -\chi = \left(\frac{\partial E}{\partial N}\right)_{\upsilon(r)} \tag{1}$$

$$\eta = \frac{1}{2}\left(\frac{\partial^2 E}{\partial N^2}\right)_{\upsilon(r)} = \frac{1}{2} = \left(\frac{\partial \mu}{\partial N}\right)_{\upsilon(r)} \tag{2}$$

Pearson and Parr were presented the operational and approximate definitions depending on electron affinity (A) and ionization energy (I) of any chemical species (atom, ion or molecule) for chemical hardness, which measures of the resistance of a chemical species to charge transfer, softness (σ) electronegativity and chemical potential in the light of finite differences method [32].

$$\chi = -\mu = \left(\frac{I + A}{2}\right) \tag{3}$$

$$\eta = \frac{I - A}{2} \tag{4}$$

The global softness [33] is defined as the inverse of the global hardness and this quantity is given as in Eq. (4).

$$\sigma = \frac{1}{\eta} \tag{5}$$

Molecular Orbital Theory and Conceptual Density Functional Theory gained a new dimension with the Koopmans theorem [34] presented in the 1930s, and to predict the ionization energy and electron affinities of chemical species, the ionization energy and electron affinity of a molecule approximate the negative values of the orbital energies of HOMO and LUMO, respectively. He predicted that it was equal. Equations (6) and (7) were obtained using Eqs. (3) and (4) to calculate hardness, electronegativity, and chemical potential with the Koopmans theorem.

$$\chi = -\mu = \left(\frac{-E_{HOMO} - E_{LUMO}}{2}\right) \tag{6}$$

$$\eta = \left(\frac{E_{LUMO} - E_{HOMO}}{2}\right) \tag{7}$$

The concept of electrophilicity (ω) as a global reactivity index similar to the chemical hardness and chemical potential has been introduced by Parr et al. [35]. This new reactivity descriptor measures the stabilization in energy when the system acquires an additional electronic charge ΔN from the environment. The electrophilicity is defined as in Eq. (8).

$$\omega = \frac{\mu^2}{2\eta} = \frac{\chi^2}{2\eta} \tag{8}$$

The global electrophilicity index (ω) is a descriptor of reactivity that allows a quantitative classification of the global electrophilic nature of a molecule within a relative scale. From the light of this index, electrophilic power of a chemical compounds is associated with its electronegativity and chemical hardness. Nucleophilicity (ε) is physically the inverse of the electrophilicity as is given in the equation below (Eq. (9)).

$$\varepsilon = 1/\omega \tag{9}$$

The solvent effect in the study was examined using the polarized continuity model (PCM) model [36].

3. Results and discussion

The experimental values of the Schiff bases in the study, 4-Amino-3-methyl-1,2,4-triazole-5-thione (AMTT) and its Derivatives (SAMTT and DBAMTT) were obtained by the study [20]. Experimentally, the inhibition activity of these bases was determined as follows: AMTT < SAMTT < DBAMTT.

Quantum chemical descriptors such as E_{HOMO}, E_{LUMO}, Energy gap ($\Delta E = E_{LUMO} - E_{HOMO}$), chemical hardness, softness, electronegativity, chemical potential, proton affinity, electrophilicity and nucleophilicity were calculated and corrosion inhibition was discussed through these parameters. Numerical values of all calculated parameters of Schiff bases and derivatives and their protonated states are given in **Tables 1–4** in gas and water solution.

We will discuss all the parameters in detail below.

3.1 Non-protonated inhibitors

According to the boundary molecular orbit theory FMO, a function of the interaction between HOMO and LUMO levels of the reacting species is defined as chemical reactivity [37]. The molecule's ability to donate electrons to a suitable acceptor with empty molecular orbitals is called E_{HOMO}, and its ability to accept electrons is called E_{LUMO}. The higher the value of the inhibitor's E_{HOMO}, the greater the inhibition efficiency and its presenting electrons to the empty d-orbit of the metal surface. The larger the molecule's ability to accept electrons depends on the lower the value of E_{LUMO} [9]. As a result of the calculations, among the molecules investigated, the lowest energy E_{HOMO}, AMTT had the lowest and DBAMTT the highest corrosion inhibition (**Table 2**). This situation is compatible with causal results. This means that the molecule that tends to adsorb the most on the metal surface is DBAMTT. SAMTT, on the other hand, has a lower number of OH groups

	E_{HOMO}	E_{LUMO}	I	A	ΔE	χ	μ	η	σ	ω	ε
MP2/3-21G* level											
AMTT	−8,04974	4,03357	8,04973	−4,03357	12,08331	2,00808	−2,00808	6,04165	0,16551	0,33371	2,99656
SAMTT	−8,06062	1,85147	8,06062	−1,85147	9,91209	3,10457	−3,10457	4,95604	0,20177	0,97238	1,0284
DBAMTT	−8,00456	2,00631	8,00456	−2,00631	10,01088	2,99912	−2,99912	5,00543	0,19978	0,89849	1,11296
HF/6–31++G**											
AMTT	−8,18144	0,67403	8,18144	−0,67403	8,85547	3,75370	−3,75370	4,42773	0,22585	1,59114	0,62848
SAMTT	−8,09001	0,85635	8,09001	−0,85635	8,94636	3,61683	−3,61683	4,47318	0,22355	1,46221	0,68389
DBAMTT	−8,01518	0,67403	8,01518	−0,67403	8,68921	3,67057	−3,67057	4,34460	0,23017	1,55006	0,64514
HF/6–311++G**											
AMTT	−8,21028	0,64518	8,21028	−0,64518	8,85547	3,78254	−3,78254	4,42773	0,22584	1,61568	0,61893
SAMTT	−8,11695	0,81961	8,11694	−0,81961	8,93656	3,64866	−3,64866	4,46828	0,2238	1,48969	0,67127
DBAMTT	−8,04266	0,65607	8,04266	−0,65607	8,69873	3,69329	−3,69329	4,34936	0,22991	1,56809	0,63771
DFT/6–31++G**											
AMTT	−5,71362	−0,94914	5,71362	0,94914	4,76447	3,33138	−3,33138	2,38224	0,41977	2,32934	0,42931
SAMTT	−5,81294	−2,06944	5,81294	2,06944	3,74349	3,94119	−3,94119	1,87175	0,53426	4,14932	0,24100
DBAMTT	−5,70954	−1,8844	5,70954	1,88440	3,82513	3,79697	−3,79697	1,91256	0,52285	3,76901	0,26532
DFT/6–311++G**											
AMTT	−5,75498	−0,93254	5,75498	0,93254	4,82244	3,34376	−3,34376	2,4112	0,41472	2,31848	0,43131
SAMTT	−5,85566	−2,10019	5,85566	2,10019	3,75547	3,97792	−3,97792	1,87773	0,53255	4,21356	0,23732
DBAMTT	−5,75308	−1,91597	5,75307	1,91596	3,83710	3,83452	−3,83452	1,91855	0,52122	3,83193	0,26096

**, ** is keyword in Gaussian software.*

Table 1.
Calculated quantum chemical parameters for non-protonated molecules in gas phase (eV).

	E_{HOMO}	E_{LUMO}	I	A	ΔE	χ	μ	η	σ	ω	ε
MP2/3-21G* level											
AMTT	−8,68975	3,72853	8,68975	−3,72853	12,41829	2,48060	−2,48060	6,20914	0,16105	0,49551	2,01811
SAMTT	−8,59261	1,69011	8,59260	−1,69011	10,28272	3,45124	−3,45124	5,14136	0,19450	1,15836	0,86328
DBAMTT	−8,44594	1,84113	8,44593	−1,84113	10,28707	3,3024	−3,3024	5,14353	0,19441	1,06015	0,94326
HF/6–31++G**											
AMTT	−8,84268	1,24302	8,84268	−1,24302	10,0857	3,79983	−3,79983	5,04285	0,19830	1,43160	0,69852
SAMTT	−8,77166	1,13853	8,77166	−1,13853	9,91019	3,81656	−3,81656	4,95509	0,20181	1,46981	0,68036
DBAMTT	−8,70771	1,14370	8,70771	−1,14370	9,85141	3,78200	−3,78200	4,92570	0,20302	1,45193	0,68874
HF/6–311++G**											
AMTT	−8,85166	1,19867	8,85166	−1,19867	10,05033	3,82649	−3,82649	5,02516	0,19899	1,45687	0,68640
SAMTT	−8,77982	1,08764	8,77982	−1,08764	9,86747	3,84608	−3,84608	4,93373	0,20268	1,49910	0,66706
DBAMTT	−8,71479	1,10070	8,71478	−1,10070	9,81549	3,80703	−3,80703	4,90774	0,20375	1,47659	0,67723
DFT/6–31++G**											
AMTT	−6,15935	−0,51811	6,15934	0,51810	5,64123	3,33873	−3,33873	2,82062	0,35453	1,97600	0,50607
SAMTT	−6,07826	−2,22754	6,07825	2,22754	3,85071	4,15290	−4,15290	1,92536	0,51938	4,47880	0,22327
DBAMTT	−6,02329	−2,04985	6,02328	2,04984	3,97344	4,03657	−4,03657	1,98672	0,50334	4,10070	0,24386
DFT/6–311++G**											
AMTT	−6,19418	−0,55294	6,19417	0,55293	5,64123	3,37355	−3,37355	2,82061	0,35453	2,01744	0,49567
SAMTT	−6,11499	−2,2621	6,11499	2,26210	3,85289	4,18854	−4,18854	1,92644	0,51909	4,55344	0,21961
DBAMTT	−6,06084	−2,08495	6,06083	2,08495	3,97588	4,07289	−4,07289	1,98794	0,50303	4,17227	0,23967

**, ** is keyword in Gaussian software.*

Table 2.
Calculated quantum chemical parameters for non-protonated molecules in aqueous solution (eV).

	E_{HOMO}	E_{LUMO}	I	A	ΔE	χ	μ	η	σ	ω	ε
MP2/3-21G* level											
AMTT	−14,5106	1,46017	14,51058	−1,46017	15,97076	6,52520	−6,52520	7,98538	0,12522	2,66601	0,37509
SAMTT	−12,0003	−1,96631	12,00031	1,96631	10,03401	6,98331	−6,98331	5,01700	0,19932	4,86013	0,20575
DBAMTT	−11,7168	−1,89665	11,71677	1,89664	9,82012	6,80671	−6,80671	4,91006	0,20366	4,71799	0,21195
HF/6–31++G**											
AMTT	−14,65644	−2,33911	14,65644	2,33911	12,31733	8,49777	−8,49777	6,15866	0,16237	5,86265	0,17057
SAMTT	−11,90371	−1,92604	11,90371	1,92604	9,97769	6,91487	−6,91487	4,98884	0,20045	4,79224	0,20867
DBAMTT	−11,7132	−1,87270	11,7132	1,87270	9,8405	6,79295	−6,79295	4,92025	0,20324	4,68921	0,21325
HF/6–311++G**											
AMTT	−14,67	−2,33775	14,670046	2,33774	12,3323	8,50389	−8,50389	6,16614	0,16217	5,86397	0,17053
SAMTT	−11,9296	−1,92413	11,929565	1,92413	10,00543	6,92684	−6,92684	5,00271	0,19989	4,79551	0,20852
DBAMTT	−11,7361	−1,87161	11,736091	1,87161	9,86447	6,80385	−6,80385	4,93223	0,20274	4,69283	0,21309
DFT/6–31++G**											
AMTT	−11,7541	−5,59797	11,75405	5,59797	6,15608	8,67601	−8,67601	3,07804	0,32488	12,22745	0,08178
SAMTT	−9,71128	−5,72859	9,71127	5,72858	3,98269	7,71993	−7,71993	1,99134	0,50217	14,96409	0,06682
DBAMTT	−9,49168	−5,48341	9,49167	5,48340	4,00826	7,48754	−7,48754	2,00413	0,49896	13,98691	0,07149
DFT/6–311++G**											
AMTT	−11,7902	−5,63498	11,790242	5,63497813	6,155264	8,71261	−8,71261	3,077632	0,324925	12,33246	0,081087
SAMTT	−9,76216	−5,76206	9,7621615	5,7620563	4,000105	7,762109	−7,762109	2,000053	0,499987	15,06219	0,066391
DBAMTT	−9,53957	−5,51824	9,5395706	5,51824036	4,02133	7,528905	−7,528905	2,010665	0,497348	14,09594	0,070942

**, ** is keyword in Gaussian software.*

Table 3.
Calculated quantum chemical parameters for protonated molecules in gas phase (eV).

	E_{HOMO}	E_{LUMO}	I	A	ΔE	χ	μ	η	σ	ω	ε
MP2/3-21G* level											
AMTT	−10,6512	2,63136	10,65116	−2,63136	13,28253	4,00990	−4,00990	6,64126	0,15057	1,21056	0,82606
SAMTT	−9,2582	1,21309	9,25820	−1,21309	10,4713	4,02255	−4,02255	5,23564	0,19099	1,54526	0,64713
DBAMTT	−9,0835	1,35350	9,08350	−1,35350	10,43701	3,865	−3,865	5,21850	0,19162	1,43127	0,69867
HF/6–31++G**											
AMTT	−10,69307	1,15159	10,69307	−1,15159	11,84466	4,77074	−4,77074	5,92233	0,16885	1,92153	0,52041
SAMTT	−9,16160	1,08275	9,16160	−1,08275	10,24435	4,03942	−4,03942	5,12217	0,19523	1,59277	0,62783
DBAMTT	−9,01874	1,08656	9,01874	−1,08656	10,1053	3,96609	−3,96609	5,05265	0,19791	1,55659	0,64242
HF/6–311++G**											
AMTT	−10,698	1,11458	10,69796	−1,11458	11,81256	4,79169	−4,79169	5,90627	0,16931	1,94372	0,51447
SAMTT	−9,18664	1,03948	9,18663	−1,03948	10,22612	4,07357	−4,07357	5,11306	0,19557	1,62271	0,61625
DBAMTT	−9,03616	1,05036	9,03615	−1,05036	10,08652	3,99289	−3,99289	5,04326	0,19828	1,58064	0,63265
DFT/6–31++G**											
AMTT	−7,79014	−1,51623	7,79013	1,51623	6,27390	4,65318	−4,65318	3,13695	0,31878	3,45113	0,28975
SAMTT	−6,90168	−2,68415	6,90167	2,68415	4,21752	4,79291	−4,79291	2,10876	0,47421	5,44680	0,18359
DBAMTT	−6,74031	−2,48605	6,74031	2,48605	4,25426	4,61318	−4,61318	2,12713	0,47011	5,00238	0,19990
DFT/6–311++G**											
AMTT	−7,81953	−1,54916	7,81952	1,54915	6,27036	4,68434	−4,68434	3,13518	0,31896	3,49948	0,28575
SAMTT	−6,95991	−2,71926	6,95991	2,71925	4,24065	4,83958	−4,83958	2,12032	0,47162	5,5231	0,18105
DBAMTT	−6,79528	−2,52224	6,79528	2,52224	4,27303	4,65876	−4,65876	2,13651	0,46805	5,07930	0,19687

**, ** is keyword in Gaussian software.*

Table 4.
Calculated quantum chemical parameters for protonated molecules in aqueous solution (eV).

than DBAMTT, so its corrosion inhibition efficiency is less than DBAMTT. **Figure 2** shows the areas where the activities of molecules are high. It is seen that these areas are the regions where nitrogen is present.

The ΔE approach, defined as the HOMO - LUMO energy deficit, is a very important stability index and provides the necessary theoretical models to make explanations about the structure and conformation in molecular systems. In order to have a high inhibition efficiency, the ΔE value should be low [38–40]. When the investigated molecules were compared, it was seen that there was a DBAMTT molecule in the smallest HOMO-LUMO gap (8,68921 eV obtained by HF method) as seen in **Table 2**. This means that the DBAMTT molecule has a tendency to adsorb more on the metal surface than other molecules and can be expected to be a very good corrosion inhibitor.

Electronegative indicates the strength of atoms in a molecule to attract bonding electrons [41]. The higher the electronegativity value of the molecule, the more the atoms in the molecule will attract the bond electrons. This will cause the inhibitor activity of the molecule to decrease [42].

The reason for not making a detailed analysis on the dipole moment μ, it should be stated first that there is no consensus on the relationship between dipole moment and inhibition efficiency in literature [7, 11, 43–46]. The results obtained in the

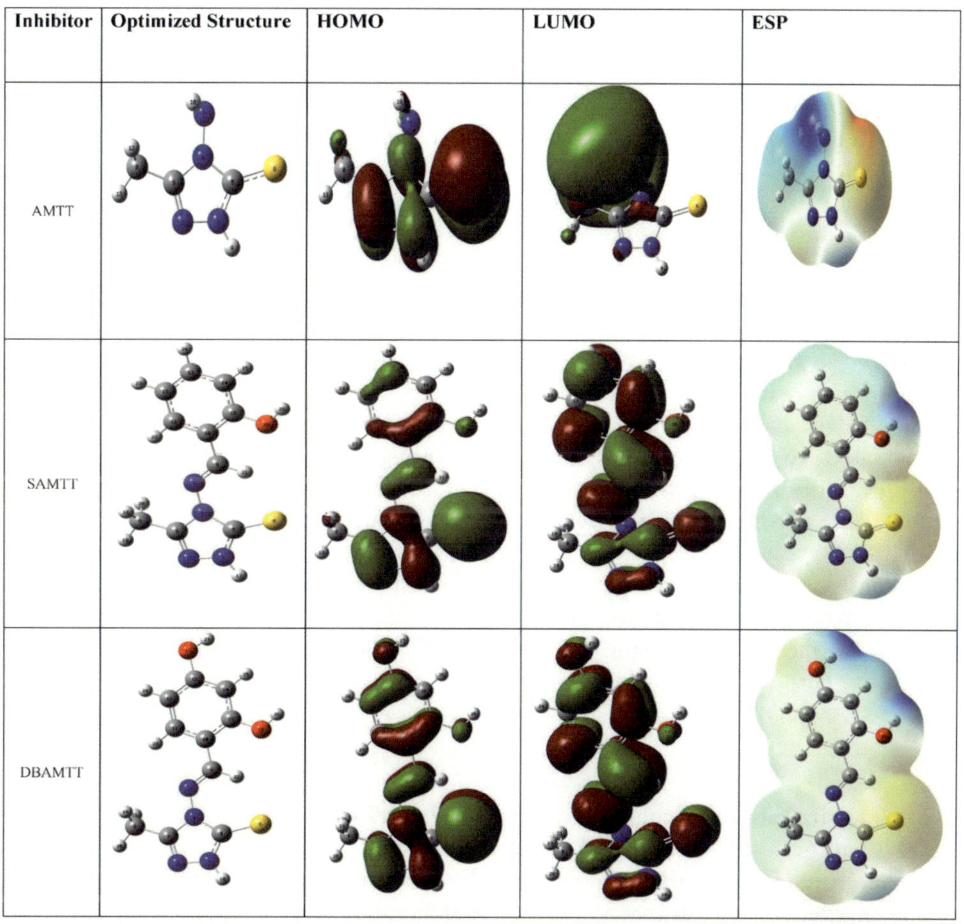

Figure 2.
*The optimized structures, HOMOs, LUMOs and electrostatic potential structures of nonprotonated inhibitor molecules using DFT/6–311++G** calculation level.*

study also show that there is no significant result between μ and inhibition efficiency and that it confirms the previous studies.

One of the ways to measure molecular stability and reactivity is Absolute hardness, η, and softness σ, where a hard molecule has a fairly large energy gap and a soft molecule has a small energy gap. Soft molecules are more reactive than hard molecules because they easily donate electrons to the receiver. While performing a simple electron transfer from the adsorption molecule, the transfer takes place from the part where the σ value of the molecule is highest [47]. In corrosion systems, the metal behaves like a lewis acid and while it is soft acid, the inhibitor acts as a lewis base and the more soft base inhibitors are, the more these metals have an effect on acidic corrosion. In this case, when **Table 2** dec-I data were examined, it was seen that the DBAMTT inhibitor had the highest σ value and this was an expected result when compared with the experimental results.

When iron and inhibitor approach each other, electrons flow from low χ (inhibitor) to high χ (iron) until their chemical potential or electronegativity is equal. As a first approximation, the fraction of electrons transferred, N, is given by Eq. (10).

$$\Delta N_{\max} = \frac{\chi_{Fe} - \chi_{inh}}{2(\eta_{Fe} + \eta_{inh})} \tag{10}$$

In the Hard Soft Acid Base (HSAB) theory [48], it was determined that iron behaves like lewis acid, so the difference in electronegativity drives the electron transfer and the sum of the hardness parameters acts as a resistance. This is calculated using the fraction of transferred electrons assuming I = A for a metallic mass, using a theoretical value χ_{Fe} = 7 eV [49] for cast iron's electronegativity and a global hardness Fe = 0. [50] In this study, the number of electrons transferred (ΔNmax) of the compounds under the probe is calculated and the results are shown in **Tables 5** and **6**. According to Lukovits et al. [51], if ΔN <3.6, inhibition efficiency increased with increasing ability to donate electrons at the metal surface. A value of ΔNmax < 3.6 eV indicates the tendency of a molecule to donate electrons to the metal surface. The results in **Tables 5** and **6** revealed that the molecules under the probe act as electron donors outside the protonated AMTT, SAMTT, and DBAMTT species in the gas phase and at the B3LYP/6–31 ++ G (d, p) theory level. Acting as an electron acceptor. The results show that the highest fraction of transferred electrons, ΔNmax, is associated with the best inhibitor (DBAMTT), while the least fraction is associated with the inhibitor with the least inhibitory activity (AMTT). In any case, the ability of inhibitor molecules to donate electrons follows the order DBAMTT> SAMTT> AMTT. These results are in good agreement with experimental studies.

Recently, according to the theory presented by Gomez et al. [38], provided that both electron transfer to the molecule and recycling from the molecule are simultaneously, the energy change and the hardness of the molecule change in direct proportion (Eq. (11)).

$$\Delta E_{back-donation} = -\frac{\eta}{4} \tag{11}$$

Eq. (11) implies that when>0 or $\Delta E_{b\text{-}d} < 0$, back-donation from molecule to metal is energetically preferred. The results reported in **Table 5** show that $\Delta E_{b\text{-}d} < 0$, therefore charge transfer to a molecule followed by re-release from the molecule is energetically favorable. Assuming that the inhibition efficiency should increase when the molecule has better adsorption on the metal surface, the inhibition efficiency should increase when the stabilization energy resulting from the interaction between the metal surface and the inhibitor increases. As expected and

	Gas phase (non-protonated)				Aqueous (non-protonated)			
	ΔN_{max}	$\Delta\psi$	$\Delta E_{b\text{-}d}$	E	ΔN_{max}	$\Delta\psi$	$\Delta E_{b\text{-}d}$	E
MP2/3-21G*								
AMTT	0,41312	−1,03114	−1,51041	−729.44927	0,36393	−0,82237	−1,55229	−729.47345
SAMTT	0,39299	−0,76545	−1,23901	−1070.53989	0,34511	−0,61237	−1,28534	−1070.55704
DBAMTT	0,39965	−0,79948	−1,25136	−1145.09747	0,35944	−0,66454	−1,28588	−1145.11973
HF/6–31++G**								
AMTT	0,36658	−0,59502	−1,10693	−732.38094	0,31729	−0,5077	−1,26071	−732.40847
SAMTT	0,37816	−0,63969	−1,11829	−1074.66910	0,32122	−0,51131	−1,23877	−1074.69232
DBAMTT	0,38316	−0,63787	−1,08615	−1149.53371	0,32665	−0,52558	−1,23143	−1149.56242
HF/6–311++G**								
AMTT	0,36332	−0,5845	−1,10693	−732.47119	0,315761	−0,50103	−1,25629	−732.49798
SAMTT	0.37501	−0,6284	−1,11707	−1074.81934	0,319627	−0,50404	−1,23343	−1074.84198
DBAMTT	0,38013	−0,6285	−1,08734	−1149.70469	0,325298	−0,51933	−1,22694	−1149.73270
DFT/6–31++G**								
AMTT	0,76999	−1,41241	−0,59556	−735.11261	0,64902	−1,18812	−0,70515	−735.135390
SAMTT	0,81709	−1,24967	−0,46794	−1079.51087	0,73937	−1,05253	−0,48134	−1079.52795
DBAMTT	0,83736	−1,34105	−0,47814	−1154.73730	0,74581	−1,10508	−0,49668	−1154.75993
DFT/6–311++G**								
AMTT	0,75817	−1,38603	−0,6028	−735.21174	0,64284	−1,16562	−0,70515	−735.23387
SAMTT	0,80471	−1,21595	−0,46943	−1079.67813	0,7297	−1,02576	−0,48161	−1079.69497
DBAMTT	0,82496	−1,3057	−0,47964	−1154.92623	0,73621	−1,07749	−0,49699	−1154.94853

**, ** is keyword in Gaussian software.*

Table 5.
Calculated quantum chemical parameters (ΔN_{max}, $\Delta\psi$, and $\Delta E_{b\text{-}d}$ (in eV) and the optimized energies (E in hartree) of the non-protonated molecules under probe in gas phase and in aqueous solution (eV).

	Gas phase (protonated)				Aqueous (protonated)			
	ΔN_{max}	$\Delta\psi$	ΔE_{b-d}	E	ΔN_{max}	$\Delta\psi$	ΔE_{b-d}	E
MP2/3-21G*								
AMTT	0,02972	−0,00706	−1,99635	−729.81437	0,22511	−0,33656	−1,66032	−729.89906
SAMTT	0,00166	−1,4E-05	−1,25425	−1070.89826	0,28434	−0,42331	−1,30891	−1070.97158
DBAMTT	0,01968	−0,0019	−1,22752	−1145.45825	0,30037	−0,47084	−1,30463	−1145.53527
HF/6–31++G**								
AMTT	−0,1216	−0,09106	−1,53967	−732.744808	0,18820	−0,20978	−1,48058	−732.83143
SAMTT	0,00853	−0,00036	−1,24721	−1075.03491	0,28899	−0,4278	−1,28054	−1075.11235
DBAMTT	0,02103	−0,00218	−1,23007	−1149.90218	0,30023	−0,45543	−1,26316	−1149.98316
HF/6–311++G**								
AMTT	−0,12195	−0,0917	−1,54154	−732.83585	0,18694	−0,20642	−1,47657	−732.922210
SAMTT	0,00731	−0,00027	−1,25068	−1075.18616	0,28617	−0,41873	−1,27826	−1075.26330
DBAMTT	0,01988	−0,00195	−1,23306	−1150.07416	0,29813	−0,44826	−1,26082	−1150.15477
DFT/6–31++G**								
AMTT	−0,27225	−0,22815	−0,76951	−735.47180	0,37406	−0,43892	−0,78424	−735.557070
SAMTT	−0,18077	−0,06507	−0,49784	−1079.87080	0,523313	−0,57750	−0,52719	−1079.94506
DBAMTT	−0,12163	−0,02965	−0,50103	−1155.10077	0,56104	−0,66955	0,53178	−1155,17815
DFT/6–311++G**								
AMTT	−0,27824	−0,23825	−0,76941	−735.57066	0,369302	−0,42759	−0,7838	−735.65562
SAMTT	−0,19052	−0,0726	−0,50001	−1080.03816	0,509454	−0,55032	−0,53008	−1080.11220
DBAMTT	−0,13153	−0,03478	−0,50267	−1155.28982	0,54791	−0,64139	−0,53413	−1155.36693

**, ** is keyword in Gaussian software.*

Table 6.
Calculated quantum chemical parameters (ΔN_{max}, $\Delta\psi$, and ΔE_{b-d} in eV) and the optimized energies (E in hartree) of the protonated molecules under probe in gas phase and in aqueous solution (eV).

in line with the experimental results [39], the calculated values of $\Delta E_{b\text{-}d}$ tend to: DBAMTT> SAMTT> AMTT.

In the theoretical studies, the ESP calculations of the molecules show the regions where the electron density is high in the molecule [40]. For this reason, molecules interact chemically by donating electrons on atoms with higher electron density. It is seen that the electron density of the sulfur atom is higher than the other atoms in the calculated molecules. For this reason, they try to be good inhibitors by interacting chemically over the molecular sulfur atom [41].

Apart from these, another important property, Sastri and Perumareddi [40] discovered by using the following equation molecule-metal interaction energy ($\Delta\psi$) can be calculated (14).

$$\Delta\psi = -\frac{(\chi_{Fe} - \chi_{inh})^2}{4(\eta_{Fe} + \eta_{inh})} \tag{12}$$

When the results are examined, molecule-metal interaction $\Delta\psi|$, is respectively as DBAMTT> SAMTT> AMTT (**Tables 5** and **6**). In addition, the initial molecule-metal interaction energy ($\Delta\psi$) order is again DBAMTT> SAMTT> AMTT.

3.2 Protonated inhibitors

The enthalpy of the reaction of a chemical species in the gas phase with the H + ion is defined as the affinity of protons (PA) [41, 52]. PA gives information about the ability of chemical compounds to donate or accept electrons and the degree of alkalinity. Compounds containing hetoroatoms such as oxygen and nitrogen tend to protonate very well in acidic environments and aqueous solutions. **Tables 7** and **8** shows the PA values of the compounds in this study with different calculation methods in gas and aqueous solution. When PA values and excision activities were compared, it was determined that the efficiency ranking was DBAMTT> SAMTT> AMTT and was consistent with the experimental result.

	MP2/3 – 21G*	HF/6–31++G**	HF/6–311++G**	DFT/6–31++G**	DFT/6–311++G**
AMTT	−2,57861	−2,54504	−2,56663	−2,41766	-2,41038
SAMTT	−2,3954	−2,59792	−2,62543	−2,43794	-2,44059
DBAMTT	−2,46101	−2,67045	−2,69756	−2,53419	−2,5375

**, ** is keyword in Gaussian software.*

Table 7.
Calculated proton affinity values of studied quinoline derivatives in gas phase using different calculation levels.

	MP2/3-21G*	HF/6–31++G**	HF/6–311++G**	DFT/6–31++G**	DFT/6–311++G**
AMTT	−4,22579	−4,15365	−4,18822	−4,11906	−4,12071
SAMTT	−3,92444	−4,07382	−4,10900	−3,99439	−3,99767
DBAMTT	−3,95166	−4,09308	−4,12942	−4,02454	−4,02952

**, ** is keyword in Gaussian software.*

Table 8.
Calculated proton affinity values of studied quinoline derivatives in aqueous solution using different calculation levels.

The following equation is used to calculate the PA values of Schiff bases compounds.

$$PA = E_{(\text{pro})} - \left(E_{(\text{non-pro})} + E_{H^{+}}\right) \quad (13)$$

In the above equation, Enon-pro and Epro are energies of non-protonated and protonated inhibitors, respectively. EH + is the energy of the H + ion and is calculated in the figure below.

In the high calculation methods selected in the study, the protonated inhibitors have lower EHOMO values compared to their non-protonated states, and the order is respectively DBAMTT, SAMTT, AMTT, and these results given in **Table 3** are consistent with the experimental inhibition efficiency.

Agreement of EHOMO values with experimental data ELUMO and T.E. and there is a correlation between these parameters and the inhibition efficiency. When the ΔE values were examined, it was seen that the lowest value belonged to DBAMTT and it was determined that it was the most intrusive (9,8405 eV).

In addition, when **Tables 2** and **3** are examined, it is seen that protonated compounds have higher μ than non-protonated compounds. Similarly, this verification was made for chemical hardness. The results show that the calculations show that non-protonated inhibitors have a more positive ΔN value than the protonated inhibitor. The DBAMTT molecule, on the other hand, has the highest ΔN value in each round, confirming that it has the same highest inhibitory properties as experimental data.

3.3 Solvent effect

The greater occurrence of the corrosion phenomenon in the solvent phase indicates that the solvent phase in the process may be important. Inhibitors may show different properties in a vacuum or in another solvent [1, 27, 41, 42]. In the study, the solvent effect on the molecular structure of the solute was calculated by the polarized continuity model (PCM) model [50]. In the PCM model, the solvent is treated as a continuous dielectric medium, and the solute is considered a molecule trapped in a cavity surrounded by the solvent. In the Gaussian 09 program, CPCM, a special version of PCM based on integral equation formalism, was used together with HF/6–31 ++ G (d, p) to examine the solver effect.). When the results are examined, a small increase is shown for the values of E_{HOMO}, E_{LUMO}, ΔE, T.E., Pi, MV, i and g, while a rather small decrease is shown for values of v and ΔN. For the molecular in this study, it was determined that the quantum chemical parameters calculated in the presence of a solvent (water) and in the gas phase did not differ significantly (**Table 5**).

3.4 Molecular dynamic simulations

Monte Carlo simulations can be used to predict interactions between inhibitor molecules and metal surface. In the study, the most stable low energy adsorption configurations of AMTT, SAMTT, DBAMTT on the Fe (110) surface were induced by Monte Carlo simulation and the configurations are shown in **Figure 3**. Outputs and descriptors, including total adsorption, solid adsorption and deformation energies are given in **Table 9**. Adsorption energy is attributed to the energy released during relaxed adsorbate components adsorbed on the substrate. Adsorption energy is the addition of solid adsorption and deformation energies of the adsorbate component. Higher values of negative adsorption energy indicate the presence of a more stable and stronger interaction between a metal and an inhibitor molecule. Monte Carlo simulation and DFT calculation results showed

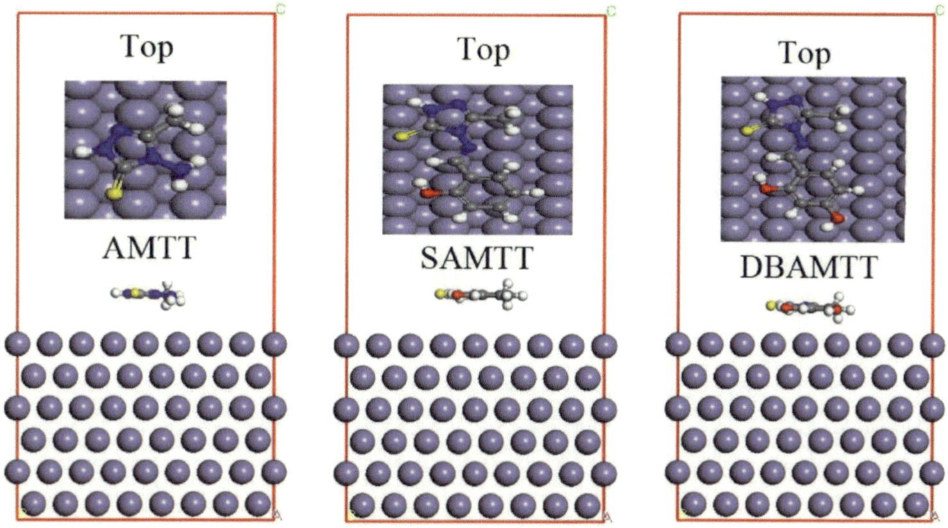

Figure 3.
Top and side views of the most stable low energy configurations for the adsorption of three inhibitors on Fe (110) interface obtained using Monte Carlo simulations.

Inhibitor	Total energy	Adsorption energy	Rigid adsorption energy	Deformation energy	dE_{ad}/dN_i	*IE* (%)
AMTT	−33,84	−164,48	−71,85	−92,63	−164,48	73
SAMTT	−172,27	−352,70	−138,95	−213,74	−352,70	93
DBAMTT	−196,85	−383,50	−143,74	−239,76	−383,50	94

Table 9.
Experimental inhibition efficiencies, IE (%) and the outputs and descriptors calculated by the Mont Carlo simulation for adsorption AMTT, SAMTT, DBAMTT of on Fe (110) (in kcal Mol^{-1}).

us once again that the corrosion inhibition efficiency was in the form of DBAMTT, SAMTT and AMTT, respectively, and it was seen to confirm the experimental results [28].

4. Conclusions

DBAMTT, SAMTT, and AMTT molecules used in this theoretical study were synthesized by M. Saravana Kumar et al. In order to predict the corrosion inhibition activities of Schiff bases and derivatives against the corrosion of iron metal, density functional theory with different basic sets and molecular dynamics simulation approach were used in Hartree Fock (HF), B3LYP. Quantum chemical calculations of the non-protonated and protonated structures of the molecules examined in this study were made in both gas phase and aqueous solution. At the end of the study, the following results are given in summary.

- The studied Schiff bases and derivatives are thought to be very important in preventing the corrosion of iron metal.

- The theoretical data obtained by both the DFT approach and the molecular dynamics simulation approach showed that the corrosion inhibition efficiency order against iron corrosion for the studied Schiff bases and derivatives can be presented as: DBAMTT> SAMTT> AMTT.

- According to the binding energies presented in **Table 8**, it was determined that among the molecules examined, DBAMTT was the most effective inhibitor of iron corrosion and the calculated binding energies were similar to the experimental data.
- Results and interpretations resulting from the study can give an idea for new corrosion inhibitor studies in the following processes and help in the selection of corrosion inhibitors.

Acknowledgements

This research was made possible by TUBITAK ULAKBIM, High Performance and Grid Computing Center (TR-Grid e-Infrastructure).

Authors Contribution

Şaban Erdoğan performed the calculations. Şaban Erdoğan and Burak Tüzün discussed and analyzed the results. All authors equally contributed to preparation of the manuscript. All the authors have read and approved the final manuscript.

Funding information

This research received no specific grant from any funding agency in the public, commercial, or not-for-profit sectors.

Compliance with ethical standards

Conflict of interest the authors declare that they have no conflict of interest.

Avaibility of data and materials

The datasets generated during the current study are available from the corresponding author on reasonable request.

Author details

Şaban Erdoğan[1*] and Burak Tüzün[2]

1 Department of Nutrition and Dietetics, Faculty of Health Sciences, Yalova University, Yalova, Turkey

2 Department of Chemistry, Faculty of Science, Cumhuriyet University, Sivas, Turkey

*Address all correspondence to: saban.erdogan@yalova.edu.tr

References

[1] N. Hackerman, Langmuir. 3 (1987) 922-924.

[2] P. Marcus, Corrosion mechanisms in theory and practice. 2002, New York Inc.: Marcel Dekker.

[3] M.S. Masoud, M.K. Awad, M.A. Shaker, M.M.T. El-Tahawy, Corr. Sci. 52 (2010) 2387-2396.

[4] C. Loganayagi, C. Kamal, M.G. Sethuraman, ACS Sustainable Chem. And Eng. 2 (2014) 606-613.

[5] I.B. Obot, S. Kaya, C. Kaya, B. Tüzün, Res. Chem. Int. 42 (2016) 4963-4983.

[6] S. Kaya, P. Banerjee, S.K. Saha, B. Tüzün, C. Kaya, RSC Advances. 6 (2016) 74550-74559.

[7] N.O. Obi-Egbedi, I.B. Obot, Mohammad I. El-Khaiary, J. Mol. Struct. 1002 (2011) 86-96.

[8] Z.S. Safi, S. Omar, Chem. Phys. Lett. 610 (2014) 321-330.

[9] S Kaya, L Guo, C Kaya, B Tüzün, IB Obot, R Touir, N Islam, Journal of the Taiwan Institute of Chemical Engineers 65 (2016) 522-529

[10] Hany M. Abd El-Lateef, Ahmed M. Abu-Dief, Mounir A.A. Mohamed, J. Mol. Struct. 1130 (2017) 522-542.

[11] S. Kaya, B. Tüzün, C. Kaya, I.B. Obot, J. Taiwan Inst. Chem. Eng. 58 (2016) 528-535.

[12] N.A. Wazzan, I.B. Obot, S. Kaya, J. Mol. Liq. 221 (2016) 579-602.

[13] S.A. Umoren, I.B. Obot, A. Madhankumar, Z.M. Gasem, Carbohydrate Polymers. 124 (2005) 280-291.

[14] Fei Yu, Shougang Chen, Yan Chen, Houmin Li, Lejiao Yang, Yuanyuan Chen, Yansheng Yin, J. Mol. Struct. 982 (2010) 152-161.

[15] Hadi Behzadi, Ali Forghani, J. Mol. Struct. 1131 (2017) 163-170.

[16] Hany M. Abd El-Lateef, Ahmed M. Abu-Dief, Mounir A.A. Mohamed, J. Mol. Struct. 1130 (2017) 522-542.

[17] M. Lebrini, M. Lagneree, H. Vezin, M. Traisnel, F. Bentiss, Corr. Sci. 49 (2007) 2254-2269.

[18] S.K. Saha, A. Dutta, P. Ghosh, D. Sukul, P. Banerjee, Phys. Chem. Chem. Phys. 17 (2015) 5679-5690.

[19] A. Dutta, S.K. Saha, P. Banerjee, D. Sukul, Corr. Sci. 98 (2015) 541-550.

[20] Kumar, M. S., Kumar, S. L. A., & Sreekanth, A. (2012). Anticorrosion potential of 4-amino-3-methyl-1, 2, 4-triazole-5-thione derivatives (SAMTT and DBAMTT) on mild steel in hydrochloric acid solution. Industrial & engineering chemistry research, 51(15), 5408-5418

[21] S. Kaya, C. Kaya, Mol. Phys. 113 (2015) 1311-1319.

[22] S. Kaya, C. Kaya, Computational and Theoretical Chemistry. 1052 (2015) 42-46.

[23] S. Kaya, C. Kaya, Computational and Theoretical Chemistry. 1060 (2015) 66-70.

[24] P.K. Chattaraj, U. Sarkar, D.R. Roy, Chem. Rev. 106 (2006) 2065-2091.

[25] H. Chermette, J. Comp. Chem. 20 (1999) 129–154.

[26] R. D. Dennington, T. A. Keith, C. M. Millam, GaussView 5.0, Wallingford CT, 2009.

[27] M.J. Frisch, G.W. Trucks, H.B. Schlegel, G.E. Scuseria, M.A. Robb, J.R. Cheeseman Jr., J.A. Montgomery, T. Vreven, K.N. Kudin, J.C. Burant, J.M. Millam, S.S. Iyengar, J. Tomasi, V. Barone, B. Mennucci, M. Cossi, G. Scalmani, N. Rega, G.A. Petersson, H. Nakatsuji, M. Hada, M. Ehara, K. Toyota, R. Fukuda, J. Hasegawa, M. Ishida, T. Nakajima, Y. Honda, O. Kitao, H. Nakai, M. Klene, X. Li, J.E. Knox, H.P. Hratchian, J.B. Cross, V. Bakken, C. Adamo, J. Jaramillo, R. Gomperts, R.E. Stratmann, O. Yazyev, A.J. Austin, R. Cammi, C. Pomelli, J.W. Ochterski, P.Y. Ayala, K. Morokuma, G.A. Voth, P. Salvador, J.J. Dannenberg, V.G. Zakrzewski, S. Dapprich, A.D. Daniels, M. C. Strain, O. Farkas, D.K. Malick, A.D. Rabuck, K. Raghavachari, J.B. Foresman, J. V. Ortiz, Q. Cui, A.G. Baboul, S. Clifford, J. Cioslowski, B.B. Stefanov, G. Liu, A. Liashenko, P. Piskorz, I. Komaromi, R.L. Martin, D.J. Fox, T. Keith, M.A. Al-Laham, C.Y. Peng, A. Nanayakkara, M. Challacombe, P.M.W. Gill, B. Johnson, W. Chen, M.W. Wong, C. Gonzalez, J.A. Pople, Gaussian 03W, Gaussian Inc., Wallingford, CT, 2004.

[28] A.D. Becke, J. Chem. Phys. 98 (1993) 5648-5652.

[29] C. Lee, W. Yang, R. Parr, Phys. Rev. 37 (1988) 785-789.

[30] K. Fukui, Angewandte Chemie International Edition in English. 21 (1982) 801-809.

[31] R.G. Parr, R.G. Pearson, J. Am. Chem. Soc. 105 (1983) 7512-7516.

[32] R.G. Pearson, Inorg. Chem. 27 (1988) 734-740.

[33] W. Yang, R.G. Parr, Proc. Natl. Acad. Sci. 82 (1985) 6723-6726.

[34] T. Koopmans, Physica 1 (1933) 104–113.

[35] R.G. Parr, L.v. Szentpaly, S. Liu, J. Am. Chem. Soc. 121 (1999) 1922-1924.

[36] I.B. Obot, S. Kaya, C. Kaya, B. Tüzün, Physica E: Low Dimensional Systems and Nanostructures. 80 (2016) 82-90.

[37] A. Popova, M. Christov, T. Deligeorgiev, Corrosion. 59 (2003) 756-764.

[38] V.R. Saliyan, A.V. Adhikari, Corr. Sci. 50 (2008) 55-61.

[39] M El Faydy, N Dahaieh, K Ounine, B Lakhrissi, I Warad, B Tüzün, A Zarrouk, Arabian Journal for Science and Engineering, (2021) 1-14

[40] A.T. Bilgiçli, H.G. Bilgicli, C. Hepokur, B. Tüzün, A. Günsel, M. Zengin, M.N. Yarasir, Applied Organometallic Chemistry, (2021) e6242

[41] E Önem, B Tüzün, S Akkoç, Journal of Biomolecular Structure and Dynamics (2021) 1-12

[42] V.S. Sastri, J.R. Perumareddi, Corrosion. 53 (1997) 617-622.

[43] G. Bereket, E. Hür, C. Öğretir, J. Mol. Struct: THEOCHEM. 578 (2002) 79-88.

[44] I.B. Obot, N.O. Obi-Egbedi, Corr. Sci. 52 (2010) 198-204.

[45] L. Guo, S. Zhu, S. Zhang, Q. He, W. Li, Corr. Sci. 87 (2014) 366-375.

[46] S. Kirkpatrick, C.D. Gelatt, M.P. Vecchi, Science 220 (1983) 671-680.

[47] R.G. Pearson, J. Am. Chem. Soc. 85 (1963) 3533-3539.

[48] S. Martinez, Mater. Chem. Phys. 77 (2003) 97-102.

[49] M.J.S. Dewar, W. Thiel, J. Am. Chem. Soc. 3533.

[50] I. Lukovits, E. Kalman, F. Zucchi, Corrosion. 57 (2001) 3-8.

[51] B. Gomez, N.V. Likhanova, M.A. Dominguez-Aguilar, R. Martinez-Palou, A. Vela, J.L. Gazquez, J. Phys. Chem. B. 110 (2006) 8928-8934.

[52] B.J. Smith, L. Radom, J. Phys. Chem. 99 (1995) 6468-6471.

Chapter 5

Corrosion and Natural Corrosion Inhibitors: A Case Study for *C. microphyllus*

Dwarika Prasad

Abstract

Worldwide, corrosion causes the value of the gross domestic product to decrease in industrialized countries by 4.26% and causes significant losses to industries including infrastructure. As a result, corrosion prevention and research related to it are extremely important. Some researchers are working to develop plant-based natural corrosion inhibitors, and experimental and computational studies are being conducted widely to prevent corrosion through cheap and environmental friendly coatings. A case study of *Convolvulus microphyllus* (*C. microphyllus*) extract was examined as eco-friendly for bio-corrosion inhibitor of mild steel in 0.5 M H_2SO_4 by using conventional weight loss, electrochemical polarization measurements, and electrochemical impedance spectroscopy (EIS) techniques. The compounds responsible for decreasing the rate of corrosion are kaempferol and phydroxycinnamic acid present in the extract. This inhibitor slows down the corrosion rate. Out of many observations, the best result 89.87% corrosion resistance efficiency was obtained at 600 mg/L of *C. microphyllus as* extract for mild steel in 0.5 M H_2SO_4 by applying electrochemical and weight loss measurements. The presence of a heteroatom in the main component of *C. microphyllus as* extract is believed to be an excellent inhibitor. Theoretical research revealed an entirely important report about comparative inhibition effect of different phytochemicals.

Keywords: adsorption, spontaneous reaction, natural inhibitors, electrochemical study, computational study

1. Introduction

The process of deposition of oxide layer on the surface of metals is called corrosion. It usually occurs when metal is exposed to moisture or water and gases such as dioxygen, dihydrogen, dichloride, hydrogen sulfide. It is a spontaneous chemical reaction with a slow rate. It usually occurs over days or weeks, and as a result of corrosion, the refined metal is turned into a more stable form such as oxide, hydroxide, or sulfide. The net worth of the corrosion prevention industry was estimated to be around $2.5 trillion (USD) in 2018 and is expected to cross $3.0 trillion (USD) by 2022, as per the National Association of Corrosion Engineers [1, 2]. The electrochemical phenomenon of degradation of material over course of time due to exposure toward the environment is called corrosion. Common types of corrosion for rusting of steel and internal polymeric pipeline are wet corrosion and dry corrosion [3]. In today's world, corrosion is no longer merely a chemical degradation of metals, but also of semiconductive materials, insulating materials, and polymeric materials after exposure to the environment. It is a surface phenomenon, where at the material surface formed oxide or hydroxide or sulfide layer [4].

Mild steel is a low-priced material with properties that are suitable for most general engineering applications. That's why it is high in demand, but it contains very poor corrosion resistance. Corrosion produces very harmful effects on commercial industries such as paper mills, oil, and gas construction, and electronics used in a multitude of processes. When materials and structures are attacked by corrosion, they lose many of their useful properties. Some useful tools and machinery made from metal can become useless because of corrosion, many disaster situations such as chemical plant leaks, bridges can collapse and oil or gas pipelines can break and can produce a dangerous effect on the life of human beings. They also produce economical losses. Acid pickling is used for removing the impurities, scale, and sludge deposits on the metal surface. Acid pickling contains strong acid due to this, that process may cause a great economic loss. In the presence of inhibitors, the rate of acidic attack decreases on the metallic surface and it prevents metallic corrosion [5]. Decades of years many organic inhibitors, for example, phosphate esters, quaternary ammonium salts, amidoethyl imidazolines, as well as an inorganic inhibitor, for example, sulfate of Mg, Mn, Ni, and Zn are being used. Corrosion of steel has been inhibited by them. These types of inhibitors are generally costly and toxic and they might be harmful to the living organisms that is why the study of eco-friendly and non-toxic green inhibitors is important to prevent steel from corrosion nowadays. Natural corrosion inhibitors are biodegradable, non-toxic eco-friendly, and low cost. Phytochemical substances obtained from plants that reduce corrosion reaction rates are termed inhibitors. The plant extract is organic in its nature, and it contains secondary-metabolized compounds such as alkaloids, amino acids, pigment, and flavonoids etc behave like inhibitors.

These types of natural products contain N, O, S, and multiple bonds so that their lone pair electrons or pi-electrons are adsorbed on the metal surface and form a protective layer to prevent metal from corrosion.

The experimental outcomes come from weight-loss study and electrochemical impedance spectroscopy studies and to support a better understanding of the adsorption of phytochemicals, computational studies were needed to operate. The quantum chemical calculations have been performed as a part of computational studies. It is well known that plant extract contains more than one phytochemical constituent. In this computational evaluation, we chose the phytochemicals present in the plant extract. The selected molecules or phytochemicals have been operated using density functional theory (DFT) using Gaussian 09 program. To decide the intensity and interaction properties of phytochemicals molecules on the metal surface Fe (110), Monte Carlo (MC), and molecular dynamics (MD) have been carried out.

2. Corrosion

Corrosion is a surface process where the oxide layer is formed on the surface of the metal. There is various types of corrosion such as oxidation corrosion, corrosion by other gases such as Cl_2, SO_2, H_2S, NO_x, liquid metal corrosion, differential metal corrosion, differential aeration corrosion, crevice corrosion, and pitting corrosion. If we discuss pitting corrosion as an example, it is due to the formation of small holes or pits on the metal surface. Pitting corrosion is highly destructive as pits are very small to be observed and usually covered with corrosion products. The pitting corrosion is the result of depassivation of a small area on the metal surface, which acts as an anode, while the rest of the undefined and large surface acts as a cathode. It is preceded by a spontaneous galvanic reaction with very limited diffusion of ions (**Figure 1**) [6].

Figure 1.
Corrosion in water pipeline.

2.1 Electrochemical aspects of corrosion

Most of the metals that we humans commonly use are unstable in the atmosphere as they were obtained from respective ores by artificial reduction through a chemical process. Thus, such metals undergo a reaction very easily to get converted into a stable form. The chemical reactions that do not require any external medium or reagent like energy or catalyst to proceed but occur on their own are called as spontaneous reactions. For spontaneous reactions, the formed products are more stable than reactants. The value for change in Gibb's free energy is always negative, and change in enthalpy is also negative, while the change in entropy is positive for such reactions.

The reactions in which oxidation and reduction both occur simultaneously are called as redox reactions. Corrosion also involves a redox reaction in which one specie or metal or part of the metal is oxidized and it acts as an anode, while the other specie or metal or part of the metal is reduced and acts as a cathode. At anode loss of electrons takes place and loss of mass occurs, while at cathode gain of electron takes place and deposition of corrosion products occurs.

The most primitive corrosion is where anodic oxidation reactions involve a pure iron when it is exposed to a strong acid such as hydrochloric acid. The reaction occurs with the formation of bubble violently. It is given as follows:

$$Fe + 2\text{H}^{+} + Cl_2^{\;-} \rightarrow Fe^{+2} + Cl^{2-} + \text{H}_2 \uparrow \quad (1)$$

$$Fe + 2\text{HCl} \rightarrow \text{FeCl}_2 + \text{H}_2 \uparrow \quad (2)$$

Another example is the exposure of iron toward moisture or water (**Figure 2**).

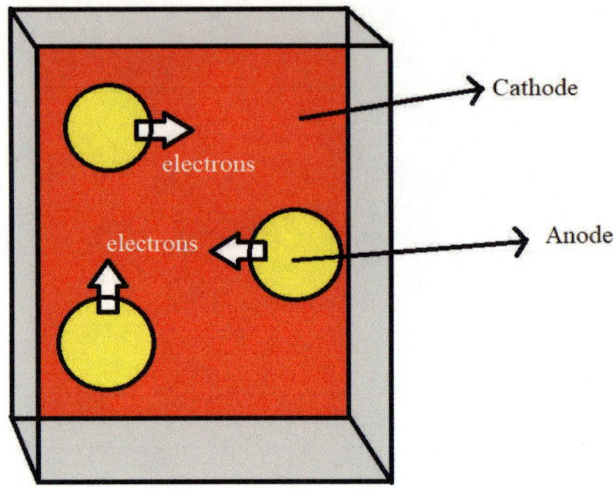

Figure 2.
Anode and cathode formation.

$$Fe \rightarrow Fe_2 + +2e^- \tag{3}$$

$$O_2 + H_2O + 4e^- \rightarrow 4OH^- \tag{4}$$

$$Fe^{2+} + 2OH^- \rightarrow Fe(OH)_2 \tag{5}$$

$$Fe^{3+} + 3OH^- \rightarrow Fe(OH)_3 \tag{6}$$

Some other reactions involve in the formation of anode and cathode that leads to corrosion are given below.

Examples of anode reactions:

$$Zn(s) \rightarrow Zn^{+2}(aq) + 2e \tag{7}$$

$$Al(s) \rightarrow Al^{+3}(aq) + 3e \tag{8}$$

$$Fe^{+2}(s) \rightarrow Fe^{+3}(aq) + e \tag{9}$$

An examples of cathode reaction:

$$H^+(aq) + e \rightarrow 1/2H_2(g) \tag{10}$$

Corrosion reactions are often electrochemical in nature. The reaction is time-consuming and may take few weeks to months to occur, it is a time and temperature-dependent reaction.

If corrosion reactions occur in aqueous media, then they are similar to that of Leclanche cell. As shown in **Figure 3**, a zinc container act as an anode, it gets oxidized. Graphite rod coated with carbon and MnO_2 paste acts as a cathode, it gets reduced, whereas both anode and cathode are joined by means of NH_4Cl and $ZnCl_2$ paste that serves the role of electrolyte. The greater the flow of electrons, the greater is the corrosion of the zinc electrode [7, 8].

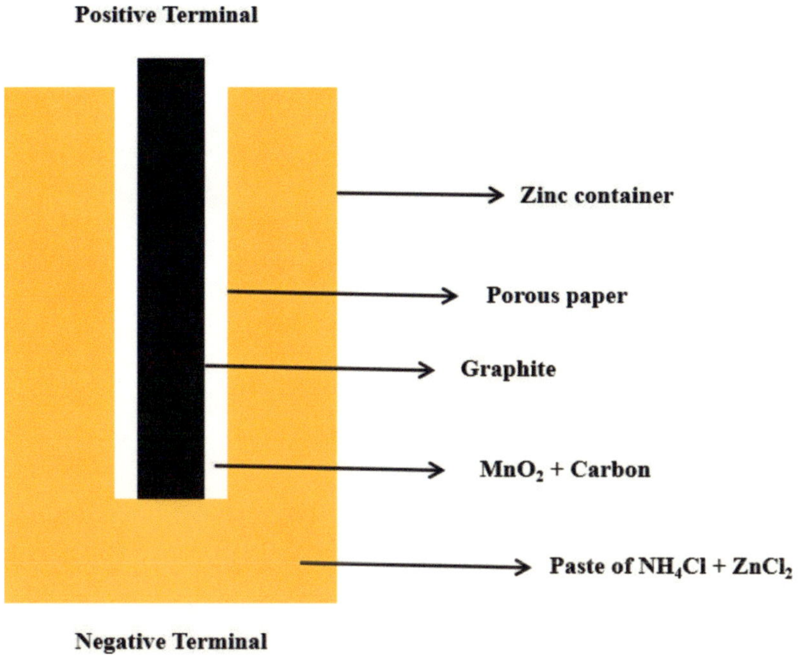

Figure 3.
Dry cell components.

2.2 Local cell formations

2.2.1 Dissimilar electrode cells

When two dissimilar metals come in direct contact with each other and an electrolytic substance is present between them, then dissimilar electrode cells are formed. It is basically a Galvanic cell, for example, we have a Daniel cell in which zinc and copper are in contact. In daily life, a copper pipe connected to a steel pipe provides an example of this type of corrosion cell. This cell is also referred to as galvanic coupling cell, the less noble metal becomes the anode whereas others act as a cathode (**Figure 4**).

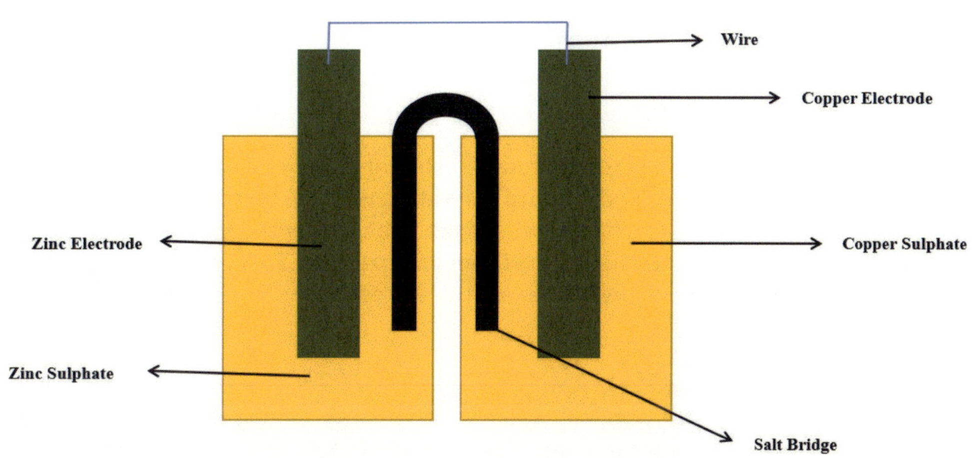

Figure 4.
Daniel cell.

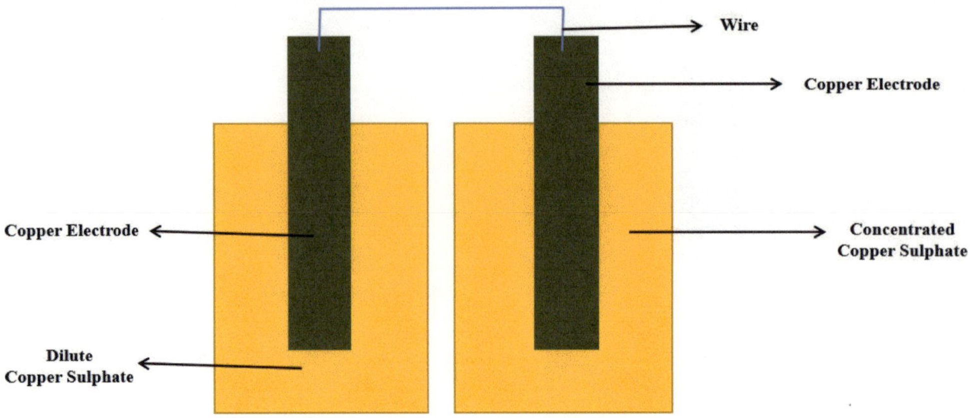

Figure 5.
Concentrated cell.

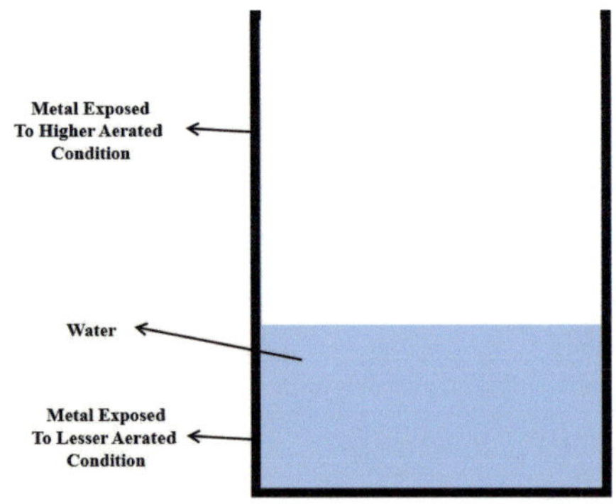

Figure 6.
Differential aeration cell.

2.2.2 Concentration cells

When two similar electrodes dipped in solutions of different compositions come in contact, then concentration cells are formed. The electrode dipped in dilute solution acts as an anode, while the electrode that is in contact with concentrated solution acts as a cathode (**Figure 5**).

2.2.3 Differential aeration cells

When two similar electrodes exposed to different aeration conditions come in contact with one another, then differential aeration cells are formed. The electrode exposed to lesser aeration conditions will act as an anode, while the electrode exposed to higher aeration conditions will act as a cathode. This type of corrosion is very common and is responsible for the significant economic loss (**Figure 6**) [9, 10].

3. Natural corrosion inhibitors

Natural corrosion inhibitors reduced or controlled the rate of corrosion by the addition of a natural product in cleaning or pickling solution. It reduces the rate of corrosion either by inhibiting oxidation at the anode or inhibiting reduction at the cathode, or both. It forms a protective layer at the metal surface, either by physical (means electrostatic attraction between natural product and metal surface) adsorption or chemisorption (means coordination bonds between natural product and metal surface).

Studies of natural corrosion inhibitors are one of the methods for protecting metal from corrosion. So here I am discussing a specific plant *C. microphyllus* as a natural corrosion inhibitor.

3.1 Experimental studies

3.1.1 Preparation of materials

The *C. microphyllus* is a hairy herb from the *Convolvuaceae* family, which is also known as Shankhpushpi and wind weed. The leaves are linear to oblong, small, and subsessile. About 1–3 flowers are produced, which are stalked. It is easily available all over the world. *C. microphyllus* contains alkaloids kaempferol, and p-hydroxycinnamic acid as its main phytochemical constituent [11, 12]. **Figure 7** shows the image of *C. microphyllus* aerial parts and its main phytochemicals.

The soxhlet apparatus is used for plant extraction. Fresh *C. microphyllus* aerial parts were collected from our surrounding and then put them in a shaded area for few days, and after drying them properly, they were grinded to convert into powder material. Now, the powder was packed in a soxhlet apparatus and refluxed for 72 hours; then, the diluted extract was collected in the round bottom flask and transformed into a gel like extract by distillation. In the next step, the gel extract is kept in the desiccators packed with silica for few days so that the extract converts

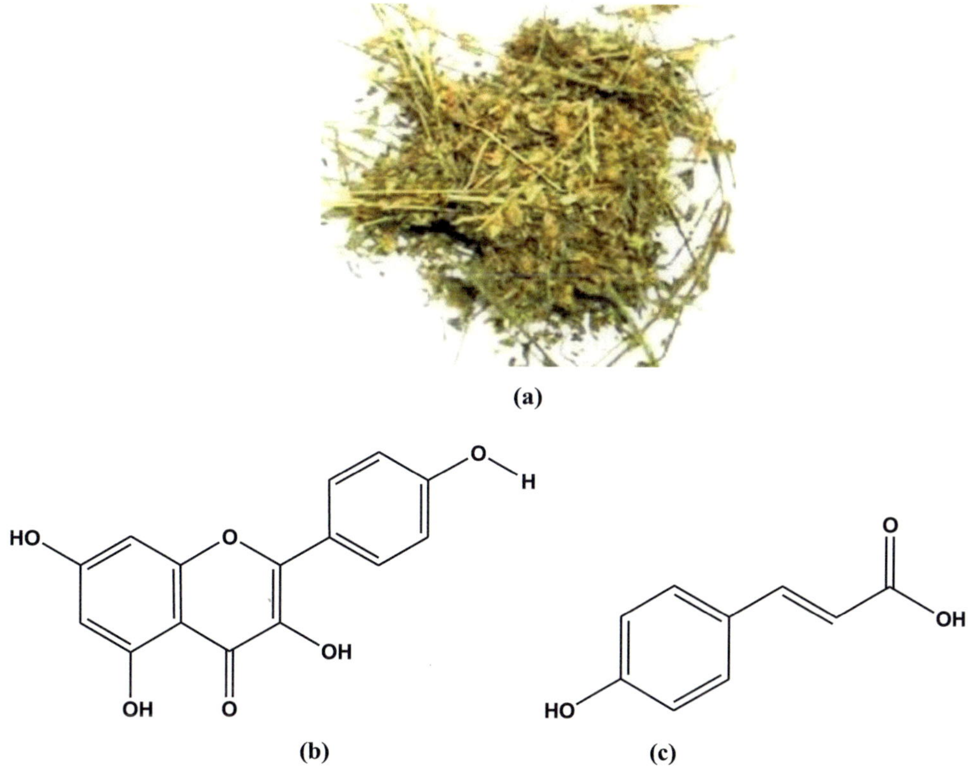

Figure 7.
(a) C. microphyllus and their main chemical components, (b) Kaempferol, and (c) p-hydroxycinnamic acid.

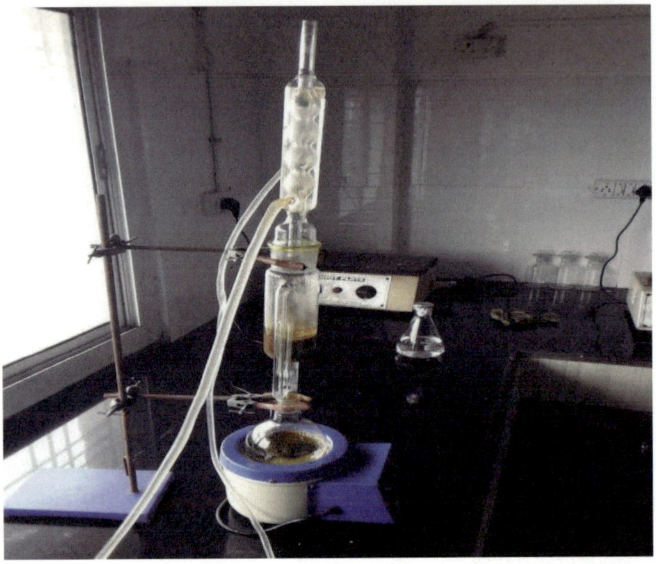

Figure 8.
Soxhlet apparatus.

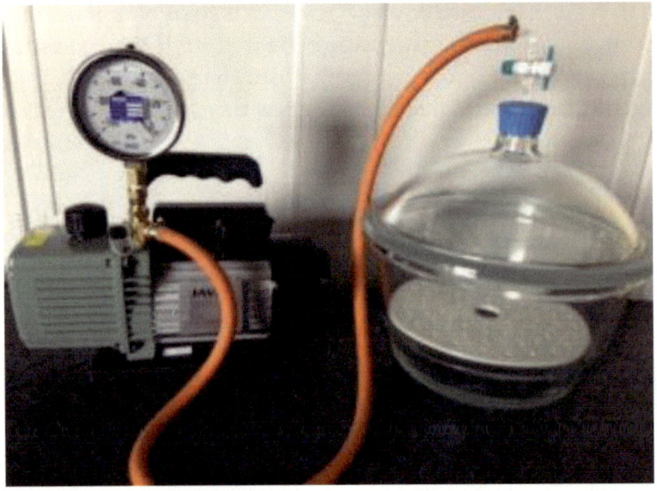

Figure 9.
Desiccators.

into solid extract without moisture. After that, it is preserved in other desiccators for experimental studies.

In this work, for weight loss measurement and electrochemical impedance studies a piece of 1 cm^2 dimension is taken as a sample. These samples were mechanically polished with emery paper of different grades and subsequently cleaned with acetone, once again clean distilled water before inserting them into the pickling or cleaning solution. Here for study of *C. microphyllus* diluted solution of conc. H_2SO_4 having a molarity of 0.5 is used as a corrosive medium (**Figures 8** and **9**).

3.1.2 Weight loss studies

The best method for this measurement is the ASTM standard G 31–72 [13]. In this method, all measurements are carried out at room temperature in the thermostat.

Inhibitor concentration (mg/L)	C_R (mmy^{-1})	Efficiency (IE %)	Surface coverage Θ
0	11.33 ± 0.12	—	—
100	3.67 ± 0.17	67.55	0.6755
200	3.08 ± 0.28	72.74	0.7274
300	2.42 ± 0.37	78.59	0.7859
400	1.89 ± 0.21	83.27	0.8327
500	1.55 ± 0.31	86.31	0.8631
600	1.35 ± 0.13	88.08	0.8808

Table 1.
Corrosion rate and efficiency of mild steel in 0.5 M H_2SO_4 absence and presence of C. microphyllus for 24 hours at room temperature.

Here minimum three times repeating the same measurement for the average value. The weight loss values, surface coverage (θ), inhibition efficiency (η%), and corrosion rate (CR) are shown in **Table 1** at different concentrations of *C. microphyllus* extract for mild steel. At different concentrations of *C. microphyllus* extract, the value of weight loss, surface coverage (θ), inhibition efficiency (η%), and corrosion rate (CR) for mild steel were calculated by equations that are as follows:

$$C_R = \frac{K \times W}{A \times t \times \rho} \tag{11}$$

$$IE(\%) = \frac{C_{R-}^0 C_R^i}{C_R^0} \times 100 \tag{12}$$

$$\theta = \frac{C_{R-}^0 C_R^i}{C_R^0} \tag{13}$$

where,

- C_R = corrosion rate (mmy^{-1})
- W = weight loss of iron alloy (g)
- $K = 8.76 \times 10^4$ (constant)
- $\rho = 7.86$ g cm^{-3} (density of Fe)
- t = immersion time (h)
- A = area of the iron alloy coupon
- θ shows surface coverage values.
- C_R^i shows the corrosion rate with inhibitor.
- C_R^0 shows the corrosion rate of the iron alloy without the inhibitor.

From the above data, it is clear that the concentration of *C. microphyllus* extract is inversely proportional to the rate of corrosion, which is used as an inhibitor.

Adsorption of active phytochemicals of *C. microphyllus* extract occurs on the surface of mild steel, which reduces the surface area available for corrosion that is why the rate of corrosion decreases. The highest inhibition efficiency 88.08% is obtained at 600 mg/L.

3.1.3 Electrochemical measurements

An electrochemical workstation is used for the electrochemical measurement. The workstation is made up of three electrodes: (a) the working electrode of metal which works like anode (b) saturated calomal electrode as the reference electrode work like cathode, and (c) platinum electrode as the counter electrode for collect current. In this measurement, metal or alloy was immersed in cleaning or pickling solutions and different concentrations of *C. microphyllus* aerial part extract were added immersed for a specific time. At room temperature, the values of the open circuit potential (OCP) were noted with respect to the reference electrode. The scanning frequency from 100 kHz to 0.01 Hz is used for recording the Nyquist plot. A 5 mV signal amplitude perturbation at OCP was considered in EIS measurement. The authentic values were taken after three-time repeat measurement.

In **Figure 10** shown Nyquist plot explained that 600 mg/L *C. microphyllus* extract provided larger radius semicircles. The diameter of the capacitive loop was enlarged by increasing *C. microphyllus* extract concentration, highest at 600 mg/L explaining also the highest inhibition effect. The radiuses of the semicircles in the case of using the *C. microphyllus* extract were larger than the blank. Because of same shapes of semicircles, it means the mechanism is not change. Calculated EIS results from the following equation provided the percentage inhibition efficiency of bio-inhibitor in the electrolytes [14]:

$$IE(\%) = \frac{R_{ct} - R_{ct}^{0}}{R_{ct}} \times 100 \qquad (14)$$

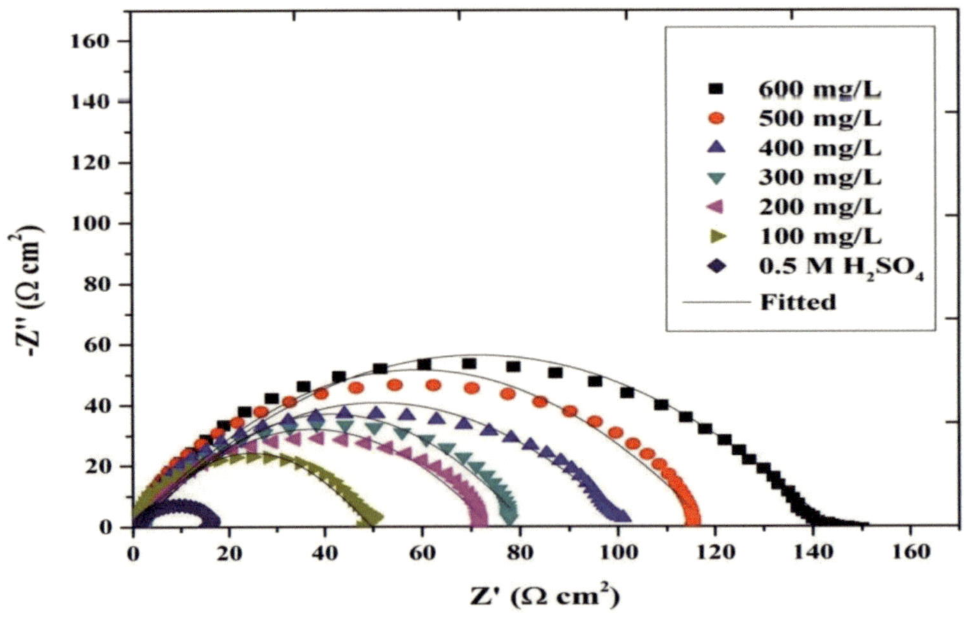

Figure 10.
Nyquist plot of mild steel in 0.5 M H_2SO_4 solution in the absence and presence of C. microphyllus extract.

Concentration of inhibitor (mg/L)	R_{ct} (Ω cm^2)	f_{max} (Hz)	R_s (Ω cm^2)	C_{dl} (μF cm^{-2})	N	Efficiency (IE %)	Θ
0	15.71	37.60	1.22	269.49	0.57	—	—
100	49.98	12.88	0.65	247.43	0.67	68.56	0.6856
200	72.58	9.46	1.63	232.19	0.68	78.35	0.7835
300	76.40	9.46	2.26	220.21	0.73	79.43	0.7943
400	104.60	7.22	0.78	211.34	0.82	84.98	0.8498
500	117.09	6.64	1.18	205.05	0.88	86.58	0.8658
600	155.13	6.64	1.68	155.15	0.91	89.87	0.8987

Table 2.
EIS parameters of mild steel in 0.5 M H_2SO_4 without and at different concentrations of C. microphyllus at 298 K.

Adsorbate	E_{HOMO}	E_{LUMO}	$\Delta E_{L\text{-}H}$	A	I	χ	H	ΔN
Hydroxycinnamic acid	−6.226	−1.996	4.230	1.996	6.226	4.111	2.115	0.1676
Kaempferol	−6.005	−2.000	4.005	2.000	6.005	4.002	2.002	0.2042

Table 3.
The HOMO and LUMO energies (eV), E_{LUMO}-E_{HOMO} energy gap ($\Delta E_{L\text{-}H}$), electron affinity (A), ionization potential (I), electronegativity (χ), hardness (η), and a fraction of electrons transferred (ΔN) for extract compounds.

where, R_{ct} and R^0_{ct} represent the charge transfer resistance in inhibitor and blank, respectively [15].

From **Table 2**, it is clear that maximum inhibition efficiency of 89.87% was obtained at 600 mg/L of *C. microphyllus* extract, where the highest R_{ct} value was 155.13 [15]. In **Table 3**, value of the double-layer capacitor (C_{dl}) gradually decreasing means protective layer gradually form at the metal surface [16].

3.2 Computational studies

3.2.1 DFT optimization

For the evaluation of electronic features of *C. microphyllus* extract molecules (including hydroxylcinnamic acid and kaempferol shown in **Figure 11**), the structures were initially optimized by density functional theory (DFT) using Gaussian 09 program. Optimization was done *via* widely used functional of B3LYP, which was combined with 6-311G** basis function. By a combination of SCRF theory and PCM formula, these calculations were continued in the liquid phase. After optimization, the following properties were analyzed: graphics, energies, and gap of molecular orbitals of highest occupied and lowest unoccupied (HOMO, LUMO, E_{HOMO}, E_{LUMO}, and E_{LUMO}-E_{HOMO} = $\Delta E_{L\text{-}H}$) and partial charges [17].

Electronic level calculations (i.e., DFT) were performed so as to obtain insights into the reactive sites of hydroxylcinnamic acid and kaempferol molecules. The optimized hydroxylcinnamic acid and kaempferol geometries resulted from chemical DFT calculations are plotted in **Figure 11**. For these energy-minimized geometries, the analysis of frontier orbitals was done as these orbitals play a crucial role in

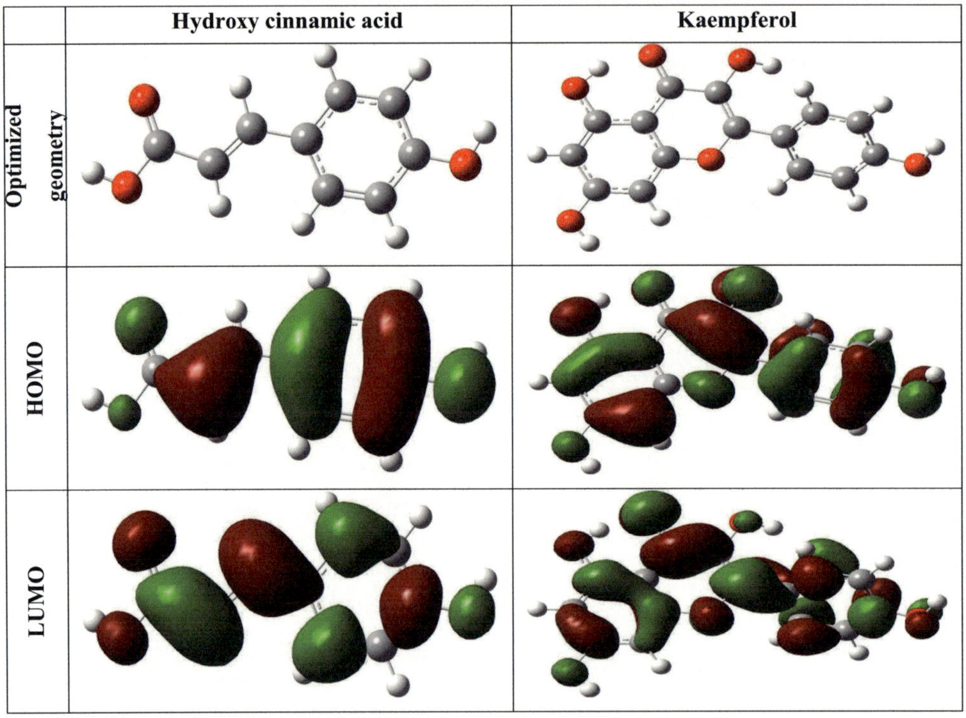

Figure 11.
*The B3LYP/6-311G** optimized geometry, HOMO, and LUMO of kaempferol and hydroxycinnamic acid compounds.*

donor-acceptor interactions of inhibiting extract molecules. The HOMO and LUMO pictures of hydroxylcinnamic acid and kaempferol are provided in **Figure 11**. As displayed, the phenyl ring, carbonyl/hydroxyl O atoms, and ethylene bond of hydroxylcinnamic acid appeared in the role of HOMO implying their strong electron supplying affinity toward a potential electron acceptor, for instance metal atoms containing unoccupied orbitals. According to the LUMO plot, all carbon and oxygen atoms of hydroxylcinnamic acid contributed to LUMO distribution, which is a signification of their tendency for getting electrons provided by filled orbitals of metal atoms. In the case of the kaempferol compound, it is seen that the HOMO placed over the entire aromatic heterocyclic skeleton and on the other hand the carbon as well as oxygen atoms behaved as reactive LUMO locations. As a consequence, the kaempferol component is able to involve in electronic interactions (i.e., donor-acceptor) with the metal surface through the donation of electrons (lone pair in hydroxyl/carbonyl O and π electrons of heterocycles) and receiving of electrons from filled iron orbitals by its C and O atoms. The eigenvalues of HOMO/LUMO and the energetic gap of frontier orbitals were examined, and the quantitative results are tabulated in **Table 3** for hydroxylcinnamic acid and kaempferol. From this table, it is evident that the kaempferol species have higher E_{HOMO} and lower E_{LUMO} in comparison with the other compound (i.e., hydroxylcinnamic acid). These results suggest the stronger electron donation and receiving capacity of kaempferol. In addition, the optimized kaempferol molecule owned a lower energetic gap (ΔE_{L-H}), an observation reflecting better electron sharing and subsequent stronger corrosion inhibition manner of this compound of *C. microphyllus* extract. This trend of DFT calculated electronic parameters and simulated adsorption energies discloses the stronger kaempferol interactions and adsorption on the steel surface leading to its more effective corrosion prevention influence as compared with the inhibiting molecule of hydroxylcinnamic acid [18, 19].

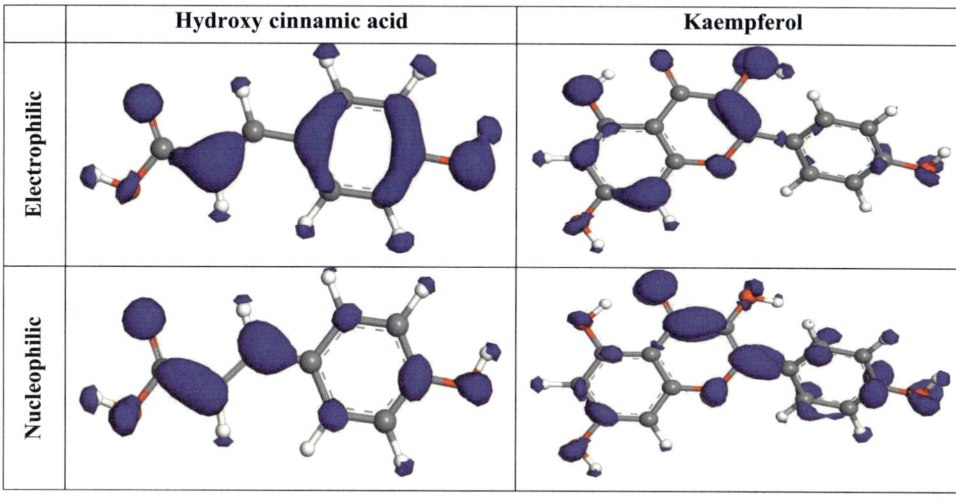

Figure 12.
The simulated Fukui indices of kaempferol and hydroxycinnamic acid compounds.

The electrophilic/nucleophilic interactions of studied molecules are summarized in **Figure 12**. From the given pictures, it is obvious that the electrophilic nature of hydroxylcinnamic acid is graphically distributed on benzene ring and double bond of carbon atoms (with a rich source of π electrons) and heteroatom of O (owing lone pair electrons). These electrons could be donated to an appropriate acceptor of electrons like metal atoms that are composed of empty orbitals. Moreover, similar to LUMO distribution, almost all carbon atoms along with three O heteroatoms are the most reactive sites for nucleophilic behavior clarifying electron getting affinity of these atomic sites. Also, the electrophilic interactions of equilibrated kaempferol compound could likely happen *via* its electron-rich places including heterocycles and the oxygen atoms of surrounding oxygenated fragments. On the other side, the O and C atoms of the whole backbone could accept electrons when attacked by a nucleophile [20, 21].

3.2.2 Molecular simulations

The hydroxylcinnamic acid and kaempferol adsorption and their interactions with the substrate (i.e., steel) were examined applying molecular simulations. These atomic simulations include Monte Carlo (MC) and molecular dynamics (MD). For the representation of steel substrate in these simulations, the frequently applied iron surface Fe (110) was used. The thickness, vacuum region, and periodic replication of this surface were set as 1.5 nm, 4 nm, and (14 × 14), respectively. To examine the adsorption preference of extract components, which are output from DFT calculations, on the iron substrate the MC-based simulation was carried out by adsorption Locator module (Materials Studio software). The MC simulations convergency was controlled with a level of fine. The last cell with the lowest potential energy generated by MC was the starting simulation box of the next modeling, that is, MD. For performing this simulation in solution phase-like experiments, water molecules were also added to the MC-generated cell. Thereafter, an optimization of 20,000 steps and NVT MD simulation of 1 ns were successively performed *via* COMPASS force field and Forcite module. The time step and temperature of NVT were set as 1 fs and 298 K. The electrostatic interactions of hydroxyl cinnamic acid and kaempferol were modeled by the Ewald scheme along with their charges calculated by DFT. The interactions of van der Waals form were evaluated by the atom-based cutoff. All metal atoms of the substrate were kept fixed [22, 23].

3.2.3 Molecular simulation

Simulations at molecular scale namely MC and MD were conducted for the analysis of the ability of *C. microphyllus* extract molecules adsorption above the metal surface (Fe (110)). The equilibrated cells of hydroxyl cinnamic acid and kaempferol generated by MC-based molecular simulations are displayed in **Figure 13**. The depicted MC snapshots clearly demonstrate that the *C. microphyllus* extract species have the ability of approaching and adsorption to iron surfaces. It is observed from the presented top views that hydroxylcinnamic acid and kaempferol adsorption on the surface happened through preferred flat orientation. The aromatic backbones of these adsorbed *C. microphyllus* extract molecules aligned parallel relative to the outmost iron atoms. The adsorption of hydroxylcinnamic acid and kaempferol took place with energies of −96.65 and − 165.51 kcal/mol, respectively. The adsorption energies of hydroxyl cinnamic acid and kaempferol with negative values further declare their adsorption over the steel substrate. The molecular adsorption was more evaluated through conducting liquid-phase dynamics (MD) simulations. The resulting snapshots of such simulations for hydroxyl cinnamic acid and kaempferol compounds are depicted in **Figure 13**. The obtained side views of ultimate simulation snapshots indicate that the chosen *C. microphyllus* extract compounds equilibrated in the vicinity of the crystalline surface (i.e., Fe (110)) under the wet conditions of the interface. This observation points to the fact that hydroxyl cinnamic acid and kaempferol as organic constituents of the green extract could bind to the surface of iron even in the presence of water molecules. Such surface adsorption affinity

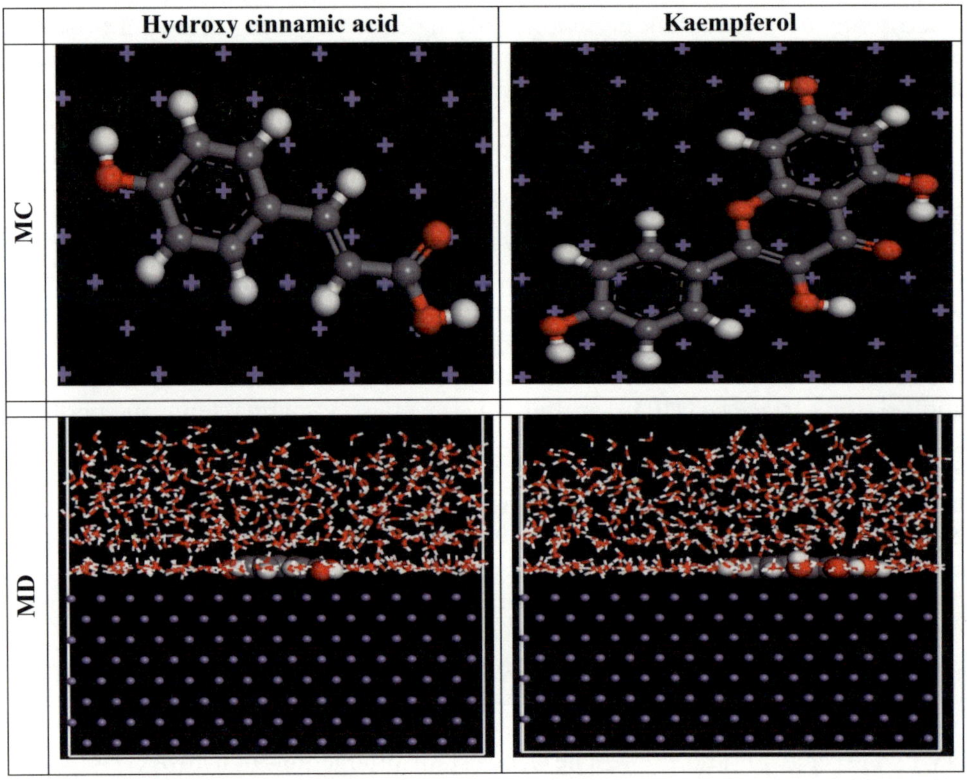

Figure 13.
The final snapshots of kaempferol and hydroxycinnamic acid compounds over Fe (110) surface obtained from MC and MD simulations.

was quantified by corresponding adsorption energy values. The energies related to hydroxyl cinnamic acid and kaempferol adsorption were estimated as −137.53 and −208.02 kcal/mol, respectively. The negative MD-calculated adsorption energies further reflect the adsorption propensity of extract components on the metallic adsorbent. In summary, the MC and MD snapshots together with predicted adsorption energies propose that the *C. microphyllus* extract compounds of hydroxyl cinnamic acid and kaempferol are capable of adsorbing on the steel substrate. The surface adsorption enables the organic extract molecules to form corrosion-preventing films over the surface, which is in support of experimental results [24, 25].

4. Conclusion

This chapter covers the corrosion problem and due to corrosion worldwide loss. This chapter discusses the importance of mild steel and discusses corrosion as a spontaneous electrochemical process. This chapter also explains of formation of different type of cells, due to different types of conditions, then start electrochemical corrosion process at the anode. In this chapter, I have discussed new developments as an example of natural corrosion inhibitor *C. microphyllus*, here I have discussed preparing inhibitors and explain experimental methods such as weight loss and electrochemical study for calculating their corrosion inhibition efficiency. After that, I have discussed computational studies of their main phytochemicals with help of DFT, molecular simulations, and molecular simulation to understand their relative adsorption properties.

Conflict of interest

The authors declare no conflict of interest.

Author details

Dwarika Prasad
Department of Chemistry, Shri Guru Ram Rai University,
Dehradun, Uttarakhand, India

*Address all correspondence to: dwarika.maithani@gmail.com

References

[1] Goyal M, Kumar S, Bahadur I, Verma C, Ebenso EE. Organic corrosion inhibitors for industrial cleaning of ferrous and non-ferrous metals in acidic solutions: A review. Journal of Molecular Liquids. 2018;**256**:565-573

[2] Verma C, Olasunkanmi LO, Ebenso EE, Quraishi MA. Substituents effect on corrosion inhibition performance of organic compounds in aggressive ionic solutions: A review. Journal of Molecular Liquids. 2018;**251**:100-118

[3] Marcus P. Corrosion Mechanisms in Theory and Practice. 3rd ed. Boca Raton, Florida: Taylor & Francis; 2011:2. DOI: 10.1201/b11020

[4] Scully JC. The Fundamentals of Corrosion. 2nd ed. USSR/Russia: International Nuclear Information System; 1978:2

[5] Haldhar R, Prasad D, Saxena A. Myristica fragrans extract as an eco-friendly corrosion inhibitor for mild steel in 0.5 M H_2SO_4. Journal of Environmental Chemical Engineering. 2018;**6**:2290-2301

[6] Fontana MG, Mitra MS. The Eight Forms of Corrosion & the Corrective Measures. India: CSIR-National Metallurgical Laboratory Jamshedpur; 1953:2-3

[7] Coleman J. Dry cell dynamics: The bobbin. Transactions of the Electro-chemical Society. 1946;**90**:545

[8] Revie RW. Corrosion and Corrosion Control: An Introduction to Corrosion Science and Engineering. Hoboken New Jersey, United States: Wiley; 2008:4

[9] Sato N. Basics of Corrosion Chemistry, Green Corrosion Chemistry. Hoboken, New Jersey United States: Wiley; 2011:6. DOI: 10.1002/9783527641789

[10] Schweitzer PA. Fundamentals of Corrosion. Boca Raton, Florida: Taylor & Francis; 2009:6. DOI: 10.1201/9781420067712

[11] Wagner H, Schwarting G. Struktur der microphyllinsaure aus dem harz von, Convolvulus microphyllus. Phytochemistry. 1977;**16**:715-717

[12] Amin H, Sharma R, Vyas M, Prajapati PK, Dhiman K. Shankhapushpi (Convolvulus pluricaulis choise): Validation of the Ayurvedic therapeutic claims through contemporary studies. International Journal of Green Pharmacy. 2014;**8**:193-200

[13] Ji G, Anjum S, Sundaram S, Prakash R. Musa paradisica peel extract as green corrosion inhibitor for iron alloyin HCl solution. Corrosion Science. 2015;**90**:107-117

[14] El-Etre AY, Abdallah M, El-Tantawy ZE. Corrosion inhibition of some metals using Lawsonia extract. Corrosion Science. 2005;**47**:385-395

[15] Joseph OO, Fayomi OSI, Joseph OO, Adenigba OA. Effect of Lecaniodiscus cupaniodes extract in corrosion inhibition of normalized and annealed mild steels in 0.5 M HCl. Energy Procedia. 2017;**119**:845-851

[16] Haldhar R, Prasad D, Bhardwaj N. Extraction and experimental studies of Citrus aurantifolia as an economical and green corrosion inhibitor for iron alloy in acidic media. Journal of Adhesion Science and Technology. 2019;**33**:1169-1183

[17] Ramezanzadeh M, Bahlakeh G, Sanaei Z, Ramezanzadeh B. Corrosion inhibition of mild steel in 1 M HCl solution by ethanolic extract of eco-friendly Mangifera indica (mango) leaves: Electrochemical, molecular dynamics, Monte Carlo and ab initio study. Applied Surface Science. 2019;**463**:1058-1077

[18] Asadi N, Ramezanzadeh M, Bahlakeh G, Ramezanzadeh B. Utilizing Lemon Balm extract as an effective green corrosion inhibitor for mild steel in 1M HCl solution: A detailed experimental, molecular dynamics, Monte Carlo and quantum mechanics study. Journal of the Taiwan Institute of Chemical Engineers. 2019;**95**:252-272

[19] Alibakhshi E, Ramezanzadeh M, Bahlakeh G, Ramezanzadeh B, Mahdavian M, Motamedi M. Glycyrrhiza glabra leaves extract as a green corrosion inhibitor for mild steel in 1 M hydrochloric acid solution: Experimental, molecular dynamics, Monte Carlo and quantum mechanics study. Journal of Molecular Liquids. 2018;**255**:185-198

[20] Sanaei Z, Ramezanzadeh M, Bahlakeh G, Ramezanzadeh B. Use of Rosa canina fruit extract as a green corrosion inhibitor for mild steel in 1M HCl solution: A complementary experimental, molecular dynamics and quantum mechanics investigation. Journal of Industrial and Engineering Chemistry. 2019;**69**:18-31

[21] Ramezanzadeh M, Sanaei Z, Bahlakeh G, Ramezanzadeh B. Highly effective inhibition of mild steel corrosion in 3.5% NaCl solution by green Nettle leaves extract and synergistic effect of eco-friendly cerium nitrate additive: Experimental, MD simulation and QM investigations. Journal of Molecular Liquids. 2018;**256**:67-83

[22] Bahlakeh G, Ramezanzadeh B, Dehghani A, Ramezanzadeh M. Novel cost-effective and high-performance green inhibitor based on aqueous Peganum harmala seed extract for mild steel corrosion in HCl solution: Detailed experimental and electronic/atomic level computational explorations. Journal of Molecular Liquids. 2019;**283**:174-195

[23] Alibakhshi E, Ramezanzadeh M, Haddadi SA, Bahlakeh G, Ramezanzadeh B, Mahdavian M. Persian Liquorice extract as a highly efficient sustainable corrosion inhibitor for mild steel in sodium chloride solution. Journal of Cleaner Production. 2019;**210**:660-672

[24] Dehghani A, Bahlakeh G, Ramezanzadeh B, Ramezanzadeh M. Detailed macro−/micro-scale exploration of the excellent active corrosion inhibition of a novel environmentally friendly green inhibitor for carbon steel in acidic environments. Journal of the Taiwan Institute of Chemical Engineers. 2019;**100**:239-261

[25] Ramezanzadeh M, Bahlakeh G, Sanaei Z, Ramezanzadeh B. Studying the Urtica dioica leaves extract inhibition effect on the mild steel corrosion in 1 M HCl solution: Complementary experimental, ab initio quantum mechanics, Monte Carlo and molecular dynamics studies. Journal of Molecular Liquids. 2018;**272**:120-136

Section 3

Characterization and Applications

Chapter 6

An Insight on Corrosion Resistance Ability of Biocompatible Dental Implants through Electrochemical Impedance Spectroscopy

Duduzile Nkomo and Nomsombuluko Masia

Abstract

Corrosion resistance is an important requirement in the study of biomedical implants. Implant surface can be modified to provide good adherence and/or optimum biocompatibility with the human body at the bone-implant interface. Titanium alloys are typically used because of their excellent corrosion resistance and biocompatibility. However, to improve these properties, the alloy surface is roughened using alumina (Al_2O_3). More details on the corrosion resistance of these alloys can be obtained by using electrochemical impedance spectroscopy (EIS) method. EIS is the most suitable method for monitoring corrosion rate values due to its reproducibility, it is non-destructive and has reliable determination of small corrosion rates, much lower than those measured by other techniques. It can also study high-impedance systems, such as coatings and linings, high-purity water, and organic coating/metal systems or corrosion in a low-conductive solution. This method has been used to evaluate electrochemical properties of modified surfaces. This chapter will explore the effectiveness of EIS in studying the corrosion behaviour of machined and surface-modified Pure Ti grade 4 for dental implant applications. The basic EIS concepts are discussed and their derivation thereof to provide information about the corrosion resistance of biomedical implants is explored.

Keywords: Biomedical, EIS, Surface modification, Implants

1. Introduction

There are multiple ways to study the corrosion resistance of biomedical implants. The common laboratory procedures include open circuit potential test (OCP), potentiodynamic polarisation test (PDP) and EIS. The commonly used electrochemical methods present measurements of currents, electrical charges or electrode potentials as a function of time. In the previous years, the use of EIS method was seldom because contrary to other electrochemical methods, EIS presents the signal as a function of frequency at a constant potential. This limits the fundamental understanding of the method because electrochemists think in terms of time and not frequency. Moreover, issues associated with understanding the EIS method are that it requires the knowledge of mathematics, Laplace and Fourier transforms, and complex numbers. However the evolution of this method has developed from using rather simple calculations and has rendered it useful in different applications such as solid materials, corrosion, passivity and batteries [1].

EIS has been used to investigate the corrosion resistance of biomedical implants. Currently we focus on the study of corrosion resistance of biomedical implants for dental applications. Different materials have been used for dental implants. Commercially pure titanium and titanium-aluminium-vanadium alloy (Ti-6Al-4 V) are most widely used biomaterials. They became popular because of their chemical resistance, mechanical strength, low toxicity, lightness and excellent biocompatibility. Ti and its alloys have the tendency to react with oxygen to form a stable oxide layer called titanium dioxide (TiO_2). This oxide layer is responsible for chemical stability and corrosion resistance of titanium [2].

The types of corrosive reactions that occur in the oral cavity are electrochemical, where the saliva plays the role. There are various types of corrosion that occurs in the oral cavity such as galvanic, fretting, pitting and uniform corrosion. Titanium implants coexist with different metallic dental superstructures and dental restorations in a patient mouth. This complexity can be linked to the occurrence of galvanic corrosion creating galvanic current, which passes through the surrounding tissue causing sharp pain. Extensive release of metal ions from a prosthetic device can result in adverse biological reactions and can lead to mechanical failure of a device itself. It is therefore important to conduct sufficient corrosion tests in order to avoid such failures and evaluate the efficiency of each corrosion test method used [2].

The advantages of using the EIS method are numerous. It provides a substantial information that can be further analysed in electrochemical properties investigations. Methods such as OCP, PDP and EIS have been used in the current study to determine the corrosion resistance of titanium pure Ti grade 4 in machined (TiGr4M) and roughened (TiGr4R) conditions in different Ringer's lactate solutions.

2. OCP

To compare the results obtained from each electrochemical testing method, we start with OCP investigations. The OCP test results shown in **Figure 1** for TiGr4 in the machined (TiGr4M) and roughened with alumina (TiGr4R) conditions are explained. OCP is the most common and simplest method used to study the film

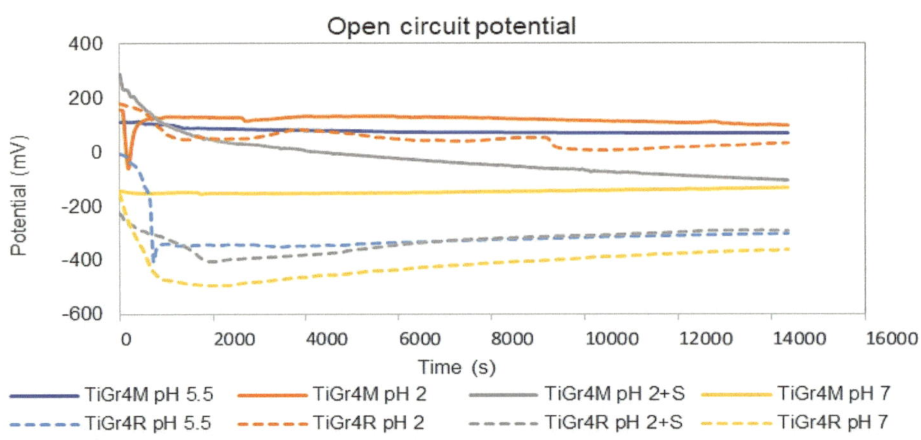

Figure 1.
OCP plots of TiGr4M and TiGr4R in different Ringer's lactate solutions.

formation and passivation of metallic structures. The potentials are measured against time and the results obtained reflect whether or not the passive film is stable in the particular environment. Thus, to mimic the nature of human mouth environment, the TiGr4M and TiGr4R samples were tested in four different pH solutions namely, the base Ringer's lactate solution-pH 5.5, Ringer's lactate-pH 2, Ringer's lactate-pH 2 with addition of sulphur and Ringer's lactate-pH 7. Each sample alloy was immersed in a solution for four hours in order to ensure stable formation of a passive oxide film. The onset of passive oxide layer stabilisation can be indicated by constant potential over a period of time.

The OCP plots for TiGr4M in all solutions was observed in the positive potential region compared to TiGr4R. In the case where no external potential applied or external disturbance such as surface defects on the test material, the formed passive oxide film remains attached to the bulk material and thus no further corrosion will be observed. A higher/more positive corrosion potential/OCP is an indication of more stable oxide film in a particular environment. Thus due to surface roughening with alumina on TiGr4R, an unstable passive oxide film formation is observed. This was observed for three different Ringer's lactate solutions except for pH 7 solution. The corrosion potential was measured at negative/more active potentials, indicating the unstable nature of the formed oxide layer. It can be deduced that the OCP test is an important step in electrochemical tests as a method studying the formation of a passive film. However, further information about the stability of the passive film is required in order to qualify new materials for use in corrosive environments such as the human mouth. This information can be obtained by applying external voltage stepwise and recording the resulting current due to anodic reaction taking place at the surface of the test material/working electrode. This has been done by use of PDP tests.

3. PDP

The PDP tests are ran such that the potential of the test sample is increased stepwise, which triggers the occurrence of either the oxidation or reduction reactions on the test material surface, and as a results the current is generated. The presentation of the potential in the function of current density (or log current density) for each measured point results in obtaining the polarisation curve. In the current presented work each scan was started from −250 mV to +1500 mV versus the corrosion potential/OCP at a scanning speed of 10 mV/min. The polarisation curve was used to determine the corrosion behaviour of the TiGr4 by Tafel slope extrapolation. The PDP plots for TiGr4M and TiGr4R in different Ringers' lactate solutions are shown in **Figure 2**.

The current density measured by Tafel slope extrapolation (cathodic and anodic reaction slope) was determined and subsequently the corrosion rates were calculated. The corrosion rates obtained were very low, less than 0.02 mm/year indicating that the material tested was in the fully passive condition. The passive behaviour of a test material can be studied by observing the increased current density as it is an indication of corrosion. Thus a material with high passive current is more prone to corrosion compared to the one with lower passive currents. Using the PDP plots, it can be observed that TiGr4R (dotted lines) showed higher passive currents than TiGr4M (solid lines) in all solutions, meaning with further exposure into Ringer's lactate solutions, the passive oxide film formed in TiGr4R has a tendency to move from passive-active as was indicated with OCP test results.

The PDP and OCP measurement are more dependent on time and thus limits these methods because with further exposure to corrosive environment, the

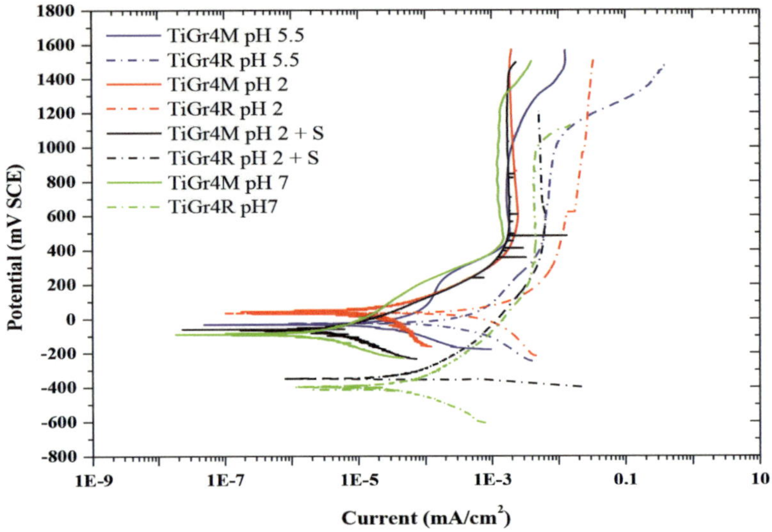

Figure 2.
Potentiodynamic polarisation plots of TiGr4M and TiGr4R in different Ringer's solutions.

exposed material can move from passive to active. Furthermore, using Tafel slope calculations for determination of corrosion rate requires that one uses data further away from the corrosion potential for accurate extrapolation and thus incorporating other reactions, which may bring significant errors in the obtained results, especially in this case where an additional phase of alumina has been introduced in TiGr4R. The EIS method can therefore be useful to obtain more accurate results in the study of electrochemical properties of this surface-modified and machined TiGr4 used for dental implant materials.

The use of EIS is highly recommended because it contains information that can be analysed at each frequency. That is, taking impedance at each potential produces series of data at different frequencies. The difference between the commonly used methods and EIS is that, although the polarisation resistance can be obtained from steady-state measurements by a potential-current plot slope, impedance plots provides more information at every point and not just at one value of polarisation potential. That is, it can be used for fine-tuning mechanisms and determining the kinetics of processes, resistances and capacitances, and it allows for the determination of real surface areas in situ. Particularly, EIS has been used for the following studies:

- Interfacial processes: redox reactions at electrodes, absorption and electrosorption, kinetics of homogeneous reactions in solution combined with redox processes, forced mass transfer

- Geometric effects: linear, spherical, cylindrical mass transfer, limited-volume electrodes, determination of solution resistance, porous electrodes

- Application in power sources: (batteries, fuel cells, supercapacitors, membranes), corrosion, coatings and paints, electrocatalytic reactions, conductive polymers, sensors and semi-conductors

EIS calculations are carried out using mathematical formulations that provide meaningful models and therefore physically relevant. The extraction of information is carried out by determination of model parameters, which directly corresponds to physical phenomena. The model parameters provides information about the measured quantity and shows a physical expected behaviour [3].

For materials, impedance spectroscopy provides possibilities to characterise the complex electrical behaviour such as resistivity, capacitive and inductive behaviour. Therefore, it is sensitive to both surface phenomena and changes of bulk properties including conduction mechanisms. It allows quantitative characterisation of corrosion processes as well as the evaluation of the anti-corrosion performance of protective coatings on corrosive metals. The use of EIS in corrosion includes detection of surface and material changes due to exposure, prediction of the lifetime of the protective passive film, evaluation of coating systems, identifying corrosion processes that lead to failures and measurement of water uptake by coatings. The EIS method follow the evolution of surface modified materials as they undergo either artificial or natural exposure to conditions that cause corrosion failure to provide information about the corrosion kinetics, which other methods do not offer [3].

Therefore, this study focuses on the use of EIS as the better method to further evaluate the corrosion resistance of TiGr4 in the machined and roughed condition. We look at the information that is obtained from EIS in comparison to that obtained from the OCP and PDP methods.

4. Nyquist plots

The first concept of EIS to tackle is the Nyquist plot. It has been established that EIS is applied to the characterisation of electrode processes and complex interfaces. Moreover, it studies the system response to the application of a periodic small amplitude of alternating current (AC) signal. Analysis of the system response contains information about the interface, its structure and reactions taking place there. The system response is generated from an electrical circuit, i.e. application of an electrical perturbation (current, potential) causes the appearance of a response. An application of a known potential $E(t)$ to a resistance R gives a current $i(t) = E(t)/R$. When the same potential is applied to the series connection of resistance R and capacitance C, the total potential difference is a sum of potential drops of each element. Taking into account that of capacitance, $E(t) = Q(t)/C$, where Q is the charge stored in a capacitor, Eq. (1) is obtained [4].

$$\mathrm{E(t) = i(t)R + \frac{Q(t)}{C} = i(t)R + \frac{1}{C}\int_0^t i(t)dt} \tag{1}$$

Eq. (1) can be solved by using what is known as Laplace transform technique. This technique is an integral transform in which a function of time $f(t)$ is transformed into new function of a parameter s called frequency, $F(s)$ according to Eq. (2) [4].

$$\mathscr{L}[f(t)] = F(s) = \int_0^\infty f(t)\exp(-st)dt \tag{2}$$

The direct application of Laplace transform to Eq. (1), given that $\mathscr{L}\left(\int_0^t i(t)dt\right) = i(s)/s$, gives Eq. (3)

$$\mathrm{E(s) = i(s)R + \frac{i(s)}{sC}} \tag{3}$$

Thus:

$$i(s) = \frac{E(s)}{\left(R + \frac{1}{sC}\right)} \quad (4)$$

The ratio of the Laplace transforms of potential and current, i.e. $E(s)/i(s)$ is expressed in units of resistance, Ω, referred to as impedance, *Z(s)* [4]:

$$Z(s) = E(s)/i(s) = R + \frac{1}{sC} \quad (5)$$

The polarisation resistance *R or* R_p value is inversely proportional to the corrosion rate. This means, the decrease in the values of R_p with increasing concentration of corrosive ions indicate the increase in corrosion rate. This interpretation can be used to support results obtained from the OCP and PDP tests in terms of the stability of the passive layer [5].

EIS has been used to explain kinetic parameters for electron transfer at the implant/saliva interface and to understand mechanism of corrosion and passivation for metallic materials in various media conditions. We use the study conducted by AlOtaibi *et al*. [2] to simplify the interpretation of Nyquist plots and the information obtained. The typical Nyquist plots obtained after immersion in artificial saliva for different exposure times is shown in **Figure 3**. The plots shown in **Figure 3** can be best fitted to the equivalent circuit model shown in **Figure 4**. The impedance results are best analysed and interpreted in terms of equivalent electrical circuit models

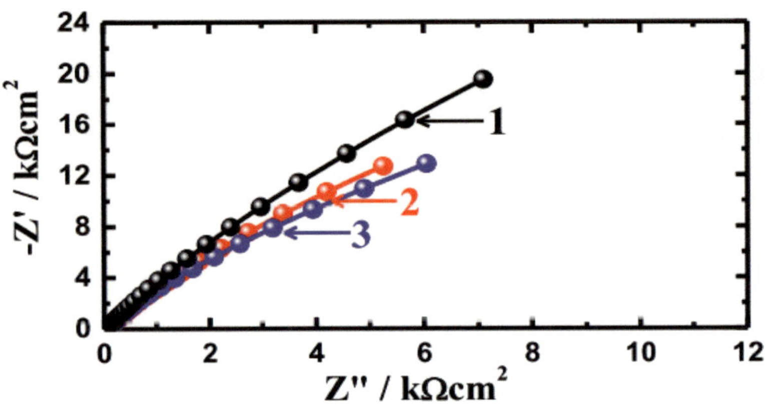

Figure 3.
Typical Nyquist plots of dental implants in an artificial saliva [2].

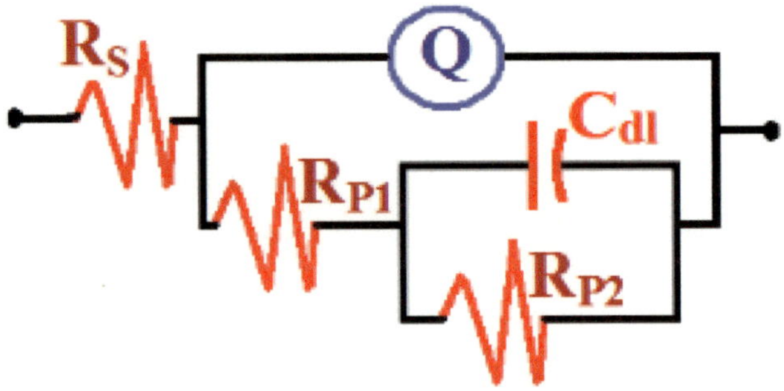

Figure 4.
The equivalent circuit model used to fit the EIS experimental data [2].

obtained by simulating the electrochemical behaviour of alloy-medium interface. The equivalent circuit refers to the presence of an oxide film layer where the inner barrier film is represented by Rp_2 and C_{d1}, but the outer layer is due to the presence of a wall capacitance, presented by Q and polarisation resistance R_{P1} of the oxide film. The overall solution resistance is presented by R_s. The electrical impedance parameters are therefore measured from this perspective.

Nyquist plots typically consists of capacitive loops at different frequencies. The loop can be at higher frequencies, medium frequencies and inductive loops at lower frequencies. The higher region frequency semicircle corresponds to the charge transfer of the corrosion process and oxide film effects. The medium frequencies semicircle corresponds to the mass transport associated with the diffusion of metallic species through the corrosion product layer. The relaxation of the surface adsorbed species is considered to be the beginning of lower frequency inductive loop. Thus, the diameter of the capacitive loops will decrease with increased concentration of corrosive species such as addition of sulphur in the Ringers solutions or decreasing of pH of the solution from 5.5 to 2 using sulphuric acid in the testing of Pure Ti dental implants [5].

We have established that Nyquist plots are often used in electrochemical literature because they allow for an easy prediction of the circuit elements. However, they do not show all details. Different types of Nyquist plots may be obtained from different values of capacitance. The only difference may be that the point on the semicircle would correspond to different frequencies. It is also possible to observe an incomplete semicircle over all the frequency range indicating a highly resistive surface [6]. This method can be used to determine the electrochemical parameters, which is AC impedance of the material. In the case of a series connection of the resistance and capacitance the impedance is given by Eq. (6). The result may be presented graphically using two plots namely Nyquist plots and Bode plots. The Nyquist plot is a plot of *Z'* (x axis) versus *Z"* (y axis), that is real versus imaginary components plotted for various frequencies.

$$Z(j\omega) = 1 + \frac{1}{j\omega C} = R - \frac{j}{\omega C} \tag{6}$$

Where: $Z(j\omega)$ stands for faradaic impedance of the circuit and ω stands for angular frequency.

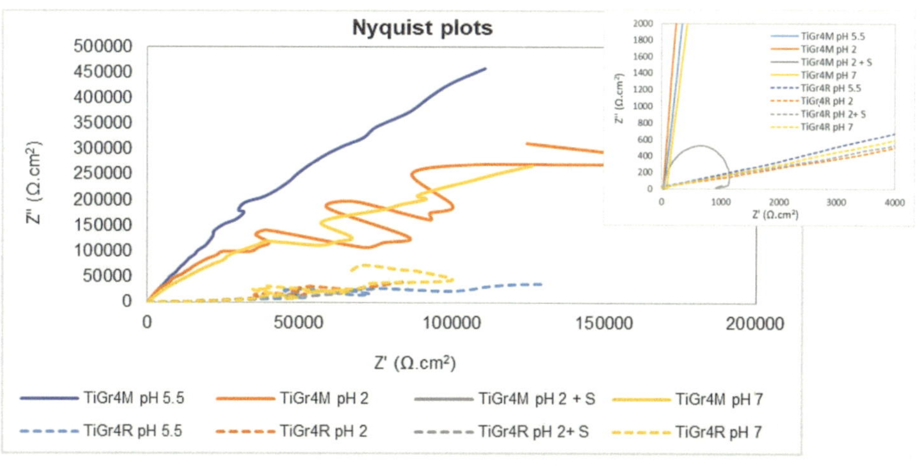

Figure 5.
Nyquist plots for TiGr4M and TiGr4R in different ringers solutions.

In our study, the EIS measurements were performed for each surface treatment of TiGr4. The frequency sweep was applied from 0.01 to 100000 Hz with an AC amplitude of 10 mA. Frequency sweep change according to material characteristics. The Nyquist plot was used to determine the capacitive behaviour, simply indicated by the size of the diameter of semicircle. **Figure 5** shows the Nyquist plots for TiGr4M and TiGr4R in different Ringer's solutions. The plots observed showed a capacitive behaviour, which is an indication of high corrosion resistance associated with incomplete semicircle observed. The increase of the diameter of the semicircle gives an indication of a high passivation of the surface against corrosion. TiGr4M showed the highest corrosion resistance indicated by the largest diameter/open incomplete semicircle. The *R* for TiGr4M is always higher than that of TiGr4R, which is in accordance with the results obtained from PDP tests where higher corrosion resistance for TiGR4M were observed.

5. Bode plots

The Bode plots have some advantages over the Nyquist plots since frequency appears as one of the axes, this shows that impedance depends on the frequency. The plots uses the logarithm of frequency to allow a very wide frequency range to be plotted on one graph, but with each decade given equal weight. The Bode plots also show the magnitude (|Z|) on a log axis so that you can easily plot wide impedance ranges on the same set of axes.

The phase angle plots are used to determine whether or not the film formed on the metal surface is intact and behaving like a capacitor. The representation shows the phase angle versus frequency (f). Frequency is generally presented on a logarithmic scale [7]. The presence of time-constant (or frequency) distribution is frequently modelled by use of a constant-phase-element (CPE). The impedance response typically reflects a distribution of reactivity that is commonly represented in equivalent electrical circuits as a CPE. Different expressions for a CPE have been presented in the literature [8], e.g.:

$$Z(\omega) = Re - \frac{j}{(j\omega To)\alpha} \tag{7}$$

If the phase is greater than 0° it exhibits a CPE behaviour. The magnitude of the complex capacitance would also then vary with the frequency. The phase angle expressed in degrees, is related to the CPE exponent n by the Eq. (8): [9].

$$\theta = 90(1 - n) \tag{8}$$

For ***n*** = 1, the CPE behaves as a capacitor; for ***n*** = 0, the CPE behaves as a resistor; and for ***n*** = −1 the CPE behaves as an inductor [8]. CPE behaviour was called frequency dispersion of capacitance because it appears as if the double-layer is behaving as a capacitor whose capacitance depends on the frequency [8, 10, 11]. Generally, the appearance of a CPE is due to the presence of inhomogeneity in the electrode material system and it can be described in terms of a distribution of relaxation times, or it may arise from non-uniform diffusion whose electrical analog is an inhomogeneous distributed RC transmission line [9].

The increased value of the phase angle at its maximum, particularly at low frequency values reveals the increased corrosion resistance [7]. **Figure 6** shows the phase angle plots for TiGr4M and TiGr4R in different Ringers solutions. Phase angles close to 80° were obtained at medium and low frequencies, suggesting a

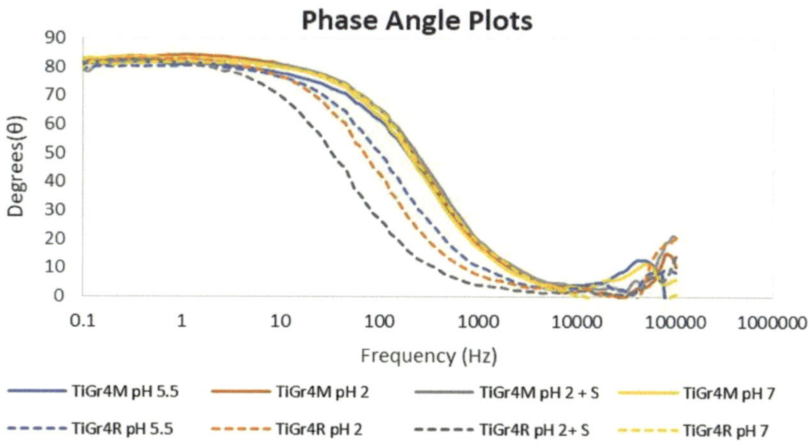

Figure 6.
Phase angle plots for TiGr4M and TiGr4R in different ringers solutions.

passive behaviour for both TiGr4M and TiGr4R. TiGr4M behaved similar in all solutions. TiGr4R showed an increased value of the phase angle at low frequencies in high pH solutions. Meaning that reducing the pH and addition of sulphide ions decreased the corrosion resistance of the TiGr4R. High values of phase angle at low frequencies for TiGr4M is always higher than that of TiGr4R, which is in accordance with the results obtained from OCP, PDP and Nyquist plots where higher corrosion resistance for TiGR4M were observed.

The modulus plot determine the dissolution kinetics of metallic surface which is represented by the magnitude of charge transfer resistance |Z|/Z'. The magnitude is usually presented on a logarithmic scale as a function of frequency on a logarithmic scale. The slope of the line at low frequencies, therefore, has a value of −1 for the blocking electrode considered [8]. The log |Z| vs. log ω curve can yield values of Rp and RΩ. At the highest frequencies, the ohmic resistance dominates the impedance and log (RΩ) can be read from the high frequency horizontal plateau. At the lowest frequencies, polarisation resistance also contributes, and log (RΩ + Rp) can be read from the low frequency horizontal plateau. At intermediate frequencies, this curve should be straight line with a slope of −1, as shown on **Figure 7**. Extrapolating this

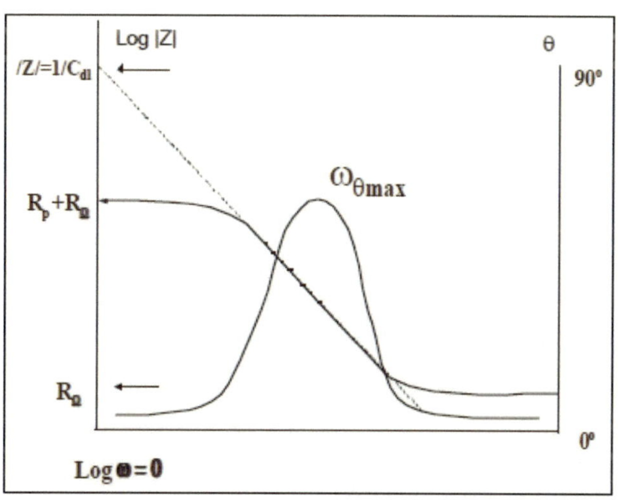

Figure 7.
Bode plot for a simple electrochemical system [12].

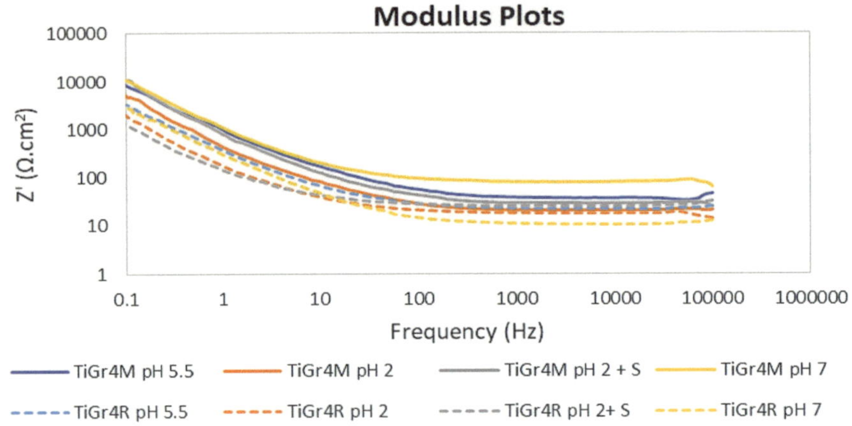

Figure 8.
Bode impedance of the interface, |Z|, plots for TiGr4M and TiGr4R in different ringers solutions.

line to the log |Z| axis at $\omega = 1$ yields the value of C_{DL} from the relationship, where $\omega = 2\pi F$ [12]:

$$|Z| = 1/C_{DL} \tag{9}$$

The higher the values of |Z|, especially at lower frequency range represents a higher corrosion resistance of the material [7]. The change of the Bode impedance of the interface (|Z|) with frequency for TiGr4M and TiGr4R is shown in **Figure 8**. The highest values of |Z|, particularly at the low frequency were recorded on both TiGr4M and TiGr4R in Ringer's pH 5.5 and Ringer's pH 7. The highest values of |Z|, at low frequencies for TiGr4M is always higher than that of TiGr4R, which is in accordance with the results obtained from PDP, Nyquist and Phase angle plots. Thus Bode impedance plots further proved higher corrosion resistance for TiGR4M compared to TiGr4R.

6. Conclusion

Corrosion testing of TiGr4 in roughened and machined conditions has been conducted. OCP tests were conducted for determination of corrosion potential of TiGr4. The passivity behaviour was studied using the PDP method. It was noted that both samples were capable of forming a passive oxide layer. From these tests it was observed that both samples were corrosion resistant, with TiGr4M demonstrating a slightly higher resistance.

A biomaterial used for dental implant application is subject to different corrosion reactions in the oral cavity and these reactions are electrochemical. Thus, we used the EIS models with a prominent parameter of frequency to forecast the lifetime of the passive layer of the proposed biomaterial. The requirement is that the implant should last a patient's lifetime once installed. Using EIS, we confirmed that indeed there is more information that could be obtained regarding the stability of the passive oxide layer formed/corrosion resistance during the exposure to artificial saliva. We used Nyquist plots, Bode plots and phase angle plots to generate information about the resistivity and capacitive behaviour of TiGr4 for both surface conditions in different pH artificial saliva solutions.

It was deduced that with regards to the passive nature of TiGr4M and TiGr4R: All the EIS plots showed that TiGr4M has the highest corrosion resistance in all the pH conditions while TiGr4R has the lowest. This was demonstrated by an increase of the diameter of the semicircle which indicated high passivation of the surface against corrosion. TiGr4M showed the highest corrosion resistance indicated by the largest diameter/open incomplete semicircle. The increased value of the phase angle at its maximum, particularly at low frequency values revealed the increased corrosion resistance. The higher the values of |Z|, especially at lower frequency range represents a higher corrosion resistance of a material.

The analysis of these EIS plots indicated that TiGr4 is a suitable material for dental implant application. Surface-modification with alumina does decrease the corrosion resistance but, insignificantly as the TiGr4R also persistently showed a comparative capacitive behaviour as with TiGr4M. Thus, it is beneficial to modify the surface with alumina as it increases the bone-implant adherence and the corrosion resistance is not utterly compromised. This offers a biocompatible material.

Acknowledgements

Authors would like to acknowledge Mintek and Southern Implants (PTY) LTD for financial assistance, sample preparation and approval to publish this work. The authors also wishes to appreciate the assistance from the Mintek colleagues especially Ms. Melanie Smit and Dr. Alain Mwamba.

Conflict of interest

The authors declare no conflict of interest.

Author details

Duduzile Nkomo* and Nomsombuluko Masia
Mintek, Randburg, South Africa

*Address all correspondence to: dudunk@mintek.co.za

References

[1] A. Lasia, *Electrochemical impedance spectroscopy and its applications*, vol. 9781461489. 2014.

[2] A. AlOtaibi, E. S. M. Sherif, S. Zinelis, and Y. S. Al Jabbari, "Corrosion behavior of two cp titanium dental implants connected by cobalt chromium metal superstructure in artificial saliva and the influence of immersion time," *Int. J. Electrochem. Sci.*, vol. 11, no. 7, pp. 5877–5890, 2016, doi: 10.20964/2016.07.77.

[3] J. J. Carroll, *Advenced Spectroscopy: Advanced applications: Battery Research, Bioimpedance, System design*. 2010.

[4] A. Lasia, *Electrochemical Impedance Spectroscopy and Its Applications Electrochemical Impedance Spectroscopy and its Applications Andrzej Lasia Département de chimie , Université de Sherbrooke , Sherbrooke*, no. July. 2017.

[5] M. G. Acharya and A. N. Shetty, "The corrosion behavior of AZ31 alloy in chloride and sulfate media – A comparative study through electrochemical investigations," *J. Magnes. Alloy.*, vol. 7, no. 1, pp. 98–112, 2019, doi: 10.1016/j.jma.2018.09.003.

[6] B. S. Gugelmin, L. S. Santos, H. D. A. Ponte, and C. E. B. Marino, "Electrochemical stability and bioactivity evaluation of Ti6Al4V surface coated with thin oxide by EIS for biomedical applications," *Mater. Res.*, vol. 18, no. 3, pp. 602–607, 2015, doi: 10.1590/1516-1439.201514.

[7] K. Habib and F. Al-Sabti, "Electrochemical impedence spectroscopy versus optical interferometry techniques during anodization of aluminium," *Opt. Lasers Eng.*, vol. 35, no. 4, pp. 225–232, 2001, doi: 10.1016/S0143-8166(01)00013-6.

[8] M.E Orazem and B. Tribollet, "Electrochemical Impedence Specroscopy: EIS", 2008.

[9] R. Srinivasan and F. Fasmin, "An introduction to Electrochemical Impedence Specroscopy", 2021.

[10] E. Barsoukov and J. R. Macdonald "Impedance Spectroscopy Theory, Experiment, and Applications", 2005.

[11] S. Tamilselvi, V. Raman and N. Rajendran, "Surface modification of titanium by chemical and thermal methods – electrochemical impedance spectroscopic studies", Corrosion Engineering, Science and Technology, vol.46, no. 4, pp. 585-591, 2011.

[12] Application Note AC-1, Subject: Basics of Electrochemical Impedance Spectroscopy, Princeton applied research.

Chapter 7

Hot Corrosion and Oxidation Behaviour of TiAl Alloys during Fabrication by Laser Powder Bed Additive Manufacturing Process

Ntebogeng Mogale, Wallace Matizamhuka and Prince Cobbinah

Abstract

This research paper summarises the practical relevance of additive manufacturing with particular attention to the latest laser powder bed fusion (L-PBF) technology. L-PBF is a promising processing technique, integrating intelligent and advanced manufacturing systems for aerospace gas turbine components. Some of the added benefits of implementing such technologies compared to traditional processing methods include the freedom to customise high complexity components and rapid prototyping. Titanium aluminide (TiAl) alloys used in harsh environmental settings of turbomachinery, such as low-pressure turbine blades, have gained much interest. TiAl alloys are deemed by researchers as replacement candidates for the heavier Ni-based superalloys due to attractive properties like high strength, creep resistance, excellent resistance to corrosion and wear at elevated temperatures. Several conventional processing technologies such as ingot metallurgy, casting, and solid-state powder sintering can also be utilised to manufacture TiAl alloys employed in high-temperature applications. This chapter focuses on compositional variations, microstructure, and processing of TiAl alloys via L-PBF. Afterward, the hot corrosion aspects of TiAl alloys, including classification, characteristics, mechanisms and preventative measures, are discussed. Oxidation behaviour, kinetics and prevention control measures such as surface and alloy modifications of TiAl alloys at high temperature are assessed. Development trends for improving the hot corrosion and oxidation resistance of TiAl alloys possibly affecting future use of TiAl alloys are identified.

Keywords: titanium aluminides, oxidation, hot corrosion, additive manufacturing, laser powder bed fusion

1. Introduction

Titanium aluminide (TiAl) is a member of group material referred to as intermetallics, consisting of various metals resulting in ordered crystallographic structures formed when the concentration of the alloy exceeds the solubility limit [1]. Properties as low density, high strength and elevated temperature properties make TiAl replacement candidates for nickel-based superalloys used in the aerospace and automotive industries [2–4]. One such alloy tried and tested by General Electric [5] for commercial turbofan engines is Ti-48Al-2Cr-2Nb. Despite the attractive high-temperature properties attained in research to date, the inherent poor ductility of TiAl at ambient temperatures remains a concern [6]. Over the past 20 years and recently, much work has been devoted to material tailoring through compositional variations and alloying aimed at improving room temperature ductility [7–11].

Phase evolution in TiAl alloys governs the mechanical and physical properties to be obtained. Primarily, two ordered structures exist, namely, γ-TiAl ($L1_0$) and hexagonal α_2-Ti_3Al ($D0_{19}$), resulting from different thermo-mechanical treatments. Furthermore, the mechanical properties to be obtained are dependent on the microstructure. Three microstructures exist, namely, equiaxed single γ phase, fully or near (γ/α_2) lamellar and duplex (consisting of colonies of lamellar γ/α_2 and pure γ phase grains). The achieved microstructure is significant for its mechanical properties, especially in structural applications. Duplex microstructures with enhanced ductility measures such as fracture strength, yield strength and strain have been reported [12–14]. Fully lamellar structures, in particular, have shown the best creep performance as contrasted to other microstructural modifications [15–17].

For the intended application, considering the inherent brittle nature of TiAl alloys, material tailoring through microstructural evolution is often necessary. Additionally, the low ductility and brittleness of TiAl alloys at ambient temperatures make their processing using conventional methods difficult. To overcome problems associated with conventional processing, such as microstructural inconsistencies inherited from solidification and phase evolutions resulting in the scattering of mechanical properties, heat treatment cycles are often designed [18–21]. Traditional methods requiring post-treatment are time-consuming, labour and capital intensive, waste a lot of start-up material, and require unnecessary production costs. Therefore, there is a need to manufacture TiAl alloy components without the above-mentioned technical deficiencies and limitations and satisfy industrial needs for component fabrication [22].

For the last decades of the 20th century [23], the Additive Manufacturing (AM) method has been employed to obtain objects by the subsequent material supply. AM mainly aims to complete a collection of traditional subtractive manufacturing practices while avoiding and limiting the need for post mechanical processing such as machining. Laser powder bed fusion (L-PBF) is an AM technique, historically referred to as Selective Laser Melting (SLM) developed by F&S Stereolithographietechnik GmbH with Fraunhofer ILT [24], where a component is manufactured by melting a powder bed in a layer-by-layer sequence employing laser beam irritation [25]. The L-PBF process is initiated by creating a 3D digital part model (usually scan data or a CAD file), followed by slicing the model into thin layers using special software. The powder bed is achieved by spreading powder onto the substrate surface. The powder bed is selectively melted through cross-sectional scanning generated from the 3D part model by applying a laser beam. After cross-section scanning, powder bed layering is achieved by sequentially adding layers one after the other repeatedly until the part is complete. Recent studies [25–29] have shown that L-PBF is an innovative and efficient process employed to manufacture TiAl alloys compared to historically employed traditional manufacturing processes such as casting [30–32], ingot metallurgy [33–35], or even solid-state powder sintering [36–38]. The benefits of L-PBF include short production cycles and cheaper production costs. Also, parts produced are of high quality and have been found to exhibit desirable performance [39].

Exploring AM technologies to improve on properties of TiAl and its alloys is essential. As such, mechanical properties like compressive and tensile ductility measures [40–42], wear resistance [43, 44], elevated temperature creep and oxidation resistance [45–48] superior to those processed by conventional means have been reported. Operation temperatures in new-generation gas turbines have fast-tracked progress in material development in the aerospace industry.

The dual combination of high temperatures and contaminant-containing aircraft environments shifts focus to hot corrosion and oxidation. Hot corrosion and oxidation can lead to catastrophic failures through material consumption at an unpredictably rapid rate. Much work has been devoted to understanding the hot corrosion and oxidation of TiAl alloys already [49–53]. As such, this research paper serves as a summary of the laser additive manufacturing of TiAl alloys. Particular attention is also given to the mechanisms, kinetics, prevention control and recent developments in hot corrosion and oxidation of TiAl alloys.

2. Titanium aluminide (TiAl) alloys: phase, microstructures and mechanical properties

2.1 Phase and microstructural evolutions

2.1.1 Phase evolutions

The three main phases of the Ti-Al system consist of various TiAl compounds, namely, γ-TiAl, α_2-Ti_3Al and $TiAl_3$ [1]. Of the three phases, only γ-TiAl and α_2-Ti_3Al have shown to be of engineering significance [54] with outstanding properties. They are lightweight and can be implemented for structural parts, automotive and elevated temperature aerospace applications. The γ-TiAl phase is a face-centred tetragonal ordered phase with an $L1_0$ structure. It consists of atomic layers at 90° to the c-axis [55] with lattice parameters a = 0.4005 nm, c = 0.4070 nm and a tetragonality ratio (c/a) of 1.02 [56, 57]. The compositional range of the γ-TiAl phase is from 48.5 to 66.0 at.% of Al. The α_2-Ti_3Al phase has a hexagonal DO_{19} structure with a compositional range from 20 to 38.2 at.% of Al.

The α_2-phase has high hydrogen and oxygen absorption rates and suffers from severe embrittlement, though it exhibits optimum high-temperature strength. The γ-phase has low gaseous absorption rates, outstanding oxidation resistance and poor room-temperature ductility. To maximise engineering benefits, dual-phase TiAl alloys consisting of $\gamma + \alpha_2$ phase are used. These alloys show excellent ductility [13, 58] at room temperatures due to the availability of refined lamellar colonies aiding γ-phase deformation [54, 59, 60]. The most known dual TiAl alloys with outstanding tensile properties are referred to as duplex alloys of the nominal (at.%) composition of Ti-(46–49) Al.

2.1.2 Microstructure-mechanical property relations

The four significant microstructures which may result in a Ti-Al system are namely, duplex (DP), near-gamma (NG), nearly lamellar (NL) and fully lamellar (FL). The obtained microstructures are greatly dependent on the processing route, Al compositional variations and thermo-mechanical treatments employed. Of the four, only fully lamellar and duplex have been considered necessary in engineering applications [54]. The evolutions (in **Figure 1**) of the microstructures mentioned above were be summarised in works by Cobbinah et al. [6] and Clemens et al. [61].

NG microstructures are obtained via thermal treatments slightly above the eutectoid temperature (T_{eu}), while DP microstructures are achieved between T_{eu} and α-transus temperatures. The thermal treatment implemented significantly

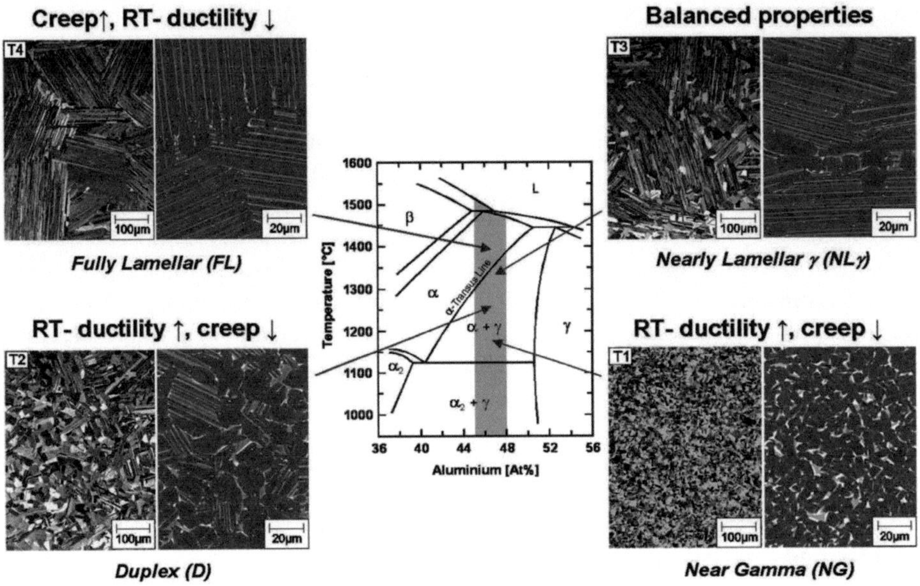

Figure 1.
The central portion of the binary Ti-Al phase diagram together with microscopic optical (left) and backscattered scanning electron (right) images showing NG, DP, NL/NLγ and FL microstructures achieved via heat-treating within α and (α + γ) phase-field. The phases obtained are identified using contrast, where a light contrast is representative of α_2-Ti_3Al and γ-TiAl of a darker contrast [61].

affects the volume fraction of lamellar grains present. As a result, NL microstructures are obtained at (T_{eu}) and T_α relative temperatures, slightly under T_α. NL microstructures exhibiting a specified globular γ-grain volume fraction are shown as NLγ. FL microstructures are achieved by thermal treatments above T_α. Generally, the obtained properties compensate for other properties [22] as represented in **Figure 1** and should be considered when the material is designed for structural applications. Furthermore, the microstructure-property relationship in TiAl alloys makes it easier to modify the material for the anticipated application.

3. Additive manufacturing (AM) of TiAl

3.1 Process overview

Additive manufacturing (AM) presents an opportunity to manufacture TiAl alloys with minimal processing difficulties compared to those experienced during conventional processing, such as near-net-shape forging or investment casting [62]. For tailoring TiAl alloys with optimum properties, laser powder bed fusion (L-PBF) and electron beam melting have been considered suitable [63–66]. Recently, the production of TiAl alloys using L-PBF has gained special attention [29, 67–71] owing to the benefits offered. Some of these benefits [6] complex geometry formation, ease of part dimension control, production of highly defined parts with orifices, mass customisation and material flexibility. Furthermore, during local melting of the powders, high solidification rates are obtained. These result in more refined microstructures.

The component is manufactured (in **Figure 2**) by melting a powder bed in a layer-by-layer sequence employing laser beam irritation [25]. The process is initiated by creating a 3D digital part model (usually scan data or a CAD file), followed by slicing the model into thin layers using special software. The powder bed is achieved by spreading powder onto the substrate surface. In preparation for part

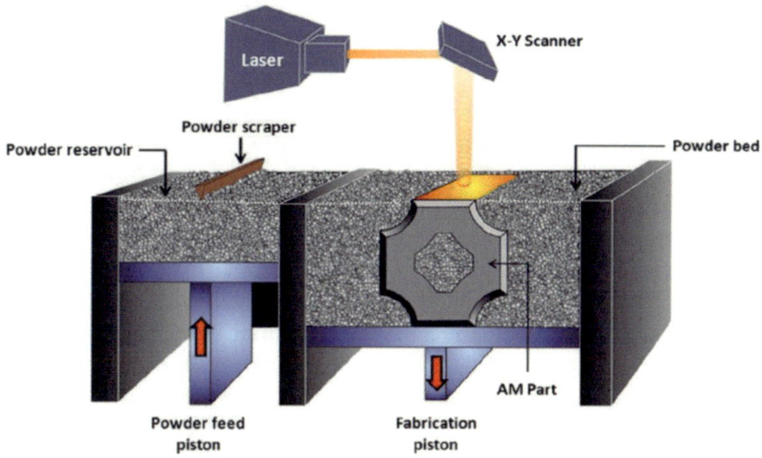

Figure 2.
Graphical representation of laser powder bed fusion method [72].

manufacturing, powders are preheated below their melting temperatures to promote bonding and minimise distortion [6]. L-PBF part manufacturing is executed in an inert gas (preferably argon) sealed environment to prevent reactive powder oxidation.

3.2 Research milestones

The need to replace previously used Ni-based superalloys in aerospace components has fast-tracked research and development of lightweight and cost-efficient TiAl alloys. To date, Ni-based alloys still outmatch TiAl alloys in fabrication costs and mechanical performance. This is mainly due to the poor room temperature ductility of TiAl alloys and the delay in engineering design practices for low ductility materials [54]. Additionally, the high part fabrication costs involved in producing TiAl alloys are related to the knowledge that low ductility fabrication processes, which also produce high melting point alloys, are unavailable. As such, there has been much investment in exploring complex part fabrication techniques, requiring minimal post-processing steps such as L-PBF.

The evidence of many research breakthroughs concerning the production of TiAl alloy parts using L-PBF does not make the processing technique immune to limitations. Efforts have been invested in overcoming processing limitations such as part cracking, micro-pore formation and uneven powder deposition through processing parameter optimisation [73]. Processing parameters can be varied to develop TiAl alloys with excellent mechanical properties in application. Some of these properties are beam size, laser power, scanning speed, scan hatch spacing and powder layer thickness [74].

Polozov et al. [75] confirmed that TiAl-based alloy crack-free samples could be built via L-PBF processing with a high-temperature platform preheating of 900°C. Fully densified samples (highest relative density of 99.9%) were attained at volume energy density 48 J/mm^3. The refined microstructure consisted of equiaxed grains, lamellar α_2/γ colonies and retained β-phase. As compared to conventionally produced TiAl alloys, high ultimate compressive strength and strain values were obtained.

Process parameters can be optimised to aid the fabrication of TiAl specimen, and unfortunately, the resultant part still shows pores, cracks and low densities. One needs to understand the crack and pore formation mechanisms and the defect-process parameter relationships in such a case. Shi and associates [70] investigated optimal L-PBF process window and the effect of substrate preheating. Moreover, the relationship between crack formation, pore formation, and the process parameters was studied and the crack propagation discrepancy with an increase in the number of deposition layers. It was concluded that crack formation was related to process parameters and the number of deposition layers. The cracks initiated in the 3rd layer are accounted for by residual stress accumulation and the deviations in the composition of Ti-47Al-2Cr-2Nb deposition layers. Furthermore, substrate (Ti-6Al-4 V) preheating at 200°C alleviated cracking. Finally, a good metallurgical bond between the substrate and Ti-47Al-2Cr-2Nb deposition layers was found.

The addition of yttrium (Y) to TiAl alloys (specifically class TNM) and process parameter optimisation dramatically affects the formability, and ultimately the cracking behaviour and control of L-PBF produced components. Gao et al. [76] fabricated TNM alloys with varying Y contents (0, 1, 2, 3, 4 wt.%) and investigated the mechanism of improved formability, cracking sensitivity, cracking behaviour and control mechanism by Y additions. Improvements in the formability of Y added-TNM alloys were assigned to lower melt viscosities and good laser energy absorption. The addition of 2, 3 and 4 wt.% Y to the TNM alloys coupled with a laser energy density greater than 7.00 J/mm^2 formed crack-free samples. The obtained microstructure and phase constituents were reported to contribute to microcrack formation and control significantly. Lower Y additions resulted in coarse columnar grains, oxygen segregation at the grain boundaries with dominating brittle B_2 phase with poor ductility. In contrast, higher Y additions (2–4 wt.%) refined equiaxed grains, enhanced the oxygen-scavenging effect (through the presence of Y_2O_3 particles), and decreased brittle B_2 phase content at higher Y additions significantly improve the ductility.

Finally, adding Nb to γ-TiAl alloys was also reported to account for improved mechanical properties based. Ismaeel et al. [77] produced Ti-Al–Mn–Nb alloys on a TC4 substrate and studied the effects of different Nb contents on the microstructure and properties of the alloys. The phases obtained consisted of γ-TiAl and α_2-Ti_3Al and a consecutive microstructural change with increased Nb additions from near full dendrite to near lamellar. Also, adding 7 at.% of Nb resulted in improved alloy's hardness, strength and plastic deformation. Moreover, the elevated temperature oxidation resistance and tribological properties were significantly improved.

4. Hot corrosion

4.1 Definition

Hot corrosion can be defined as a chemical degradation on the metallic surface of materials operating at high temperatures, enhanced by the presence of molten ash and gases containing elements such as sulphur (S), chlorine and sodium [78]. Such environmental elements during fuel combustion promote damage to the protective oxide film by forming contaminants such as V_2O_5 and Na_2SO_4 [79]. This degradation form was initially identified in the early 1950s on combustion engines and boilers [80] and has been explored in numerous research works [50, 81–87].

4.2 Characteristics

Hot corrosion exists as Type I (known as High-Temperature Hot Corrosion) or Type II (Low-Temperature Hot Corrosion), with the former occurring above 800–950°C and the latter at 600–750°C [88, 89]. The occurrence of either attack form is dependent on several parameters such as the composition of the alloy, contaminant, and gas. Furthermore, other vital parameters are temperature and temperature cycles, erosion processes and gaseous velocity [90, 91]. The main difference between high-temperature hot corrosion (HTHT) and low-temperature hot corrosion (LTHC) is the morphologies thereof. HTHC is distinguished by the occurrence of a non-porous protective scale, internal sulphidisation and chromium (Cr) depletion.

4.2.1 High-temperature hot corrosion (HTHC)-type I

This form of attack, also referred to as molten salt-induced corrosion, comprises a liquid-phase salt mixture deposit observed at high temperatures at the start of deposition [92]. Traditionally, according to Nicholls and Simms [93], HTHC has been detected in a temperature array between the surface deposit melting point and vapour deposition dew point for the deposit. Above this suggested temperature band, instability of dew point deposit exists, resulting in evaporation. A series of chemical reactions occur, initially attacking the oxide film and progress to deplete Cr present in the substrate [94]. Oxidation of the base material is then accelerated by Cr depletion, promoting a porous oxide scale formation.

An example of this could be the formation of thermodynamically unstable liquid sodium sulfate (Na_2SO_4) deposits. The marine environment mainly sources such deposits in sea salt form, followed by atmospheric contaminants such as volcanic discharges and fuel. During combustion, the present Na_2SO_4 can combine with pollutants present in air or fuel (such as chlorides, V and Pb) to form a blend of low melting temperature salts, further broadening the temperature range attack [94]. In the presence of sodium chloride (NaCl), the following reaction after combustion can be observed:

$$2NaCl + SO_2 + O_2 = Na_2SO_4 + Cl_2 \quad (1)$$

HTHC can be classified into four stages from initiation up to failure [95]:

1. Stage I: Initial coating deterioration—roughening of the surface edges coupled with localised oxide layer disintegration and minor base metal layer depletion is observed. If the surface is left untreated, the condition will worsen. Surface recoating and stripping may be adequate to remedy this degree of damage.

2. Stage II: Oxide layer rapture—characterised by an acceleration and advancement in surface roughness compared to Stage 1 and the protective oxide layer's failure. Although the mechanical integrity remains maintained, there is no way to salvage the component to its original state.

3. Stage III: Detrimental sulphidisation—depicted by massive scale build-up on the component's surface and indications of liquid Na_2SO_4 under the protective layer. The structural integrity of the part is significantly affected, attack by S contaminants proceeds.

4. Stage IV: Catastrophic attack—failure of the component occurs due to the observed significant blistered scale penetrating much into the base metal. Structural rigidity is lost.

This corrosion damage is characterised by a uniform attack, internal sulphide phases, depletion zone beneath a relatively smooth scale–metal interface [80, 96].

4.2.2 Low-temperature hot corrosion (LTHC)-type II

Type II corrosion has been reported [97–102] as a liquid-phase deterioration by a blend of molten nickel (Ni) or cobalt (Co)-containing sulphates such as Na_2SO_4-$CoSO_4$ or Na_2SO_4-$NiSO_4$ accountable for corrosion initiation and propagation. The corrosion initiation is achieved through oxide layer fluxing, while propagation is accelerated by the mass movement of reactive elemental components through liquids present in the corrosion pits [80]. Studies [103–106] have shown that conversion from CoO and NiO occurs when SO_3 in the gas reacts with the sulphates, attributing to the extensive usage of mixed Na_2SO_4-$NiSO_4$ in recent LTHC research studies.

LTHC can be found in coated or uncoated compressor and turbine parts. For instance, the sometimes turbine blade's uncoated internal cooling systems operating at temperatures of about 650–750°C may be prone to this corrosion type [107]. The external rim of uncoated turbine blades reaches temperatures of 400–800°C [108]. LTHC is distinguished by the pit's appearance and the absence of a sulphide zone at the corrosion front, consuming all the S [96].

4.3 Mechanisms

Two HTHC mechanisms have been proposed, namely sulphide-oxidation and salt fluxing mechanisms [94]. Acidic and basic fluxing reactions, presented initially by Goebel and Pettit [109, 110], may be obtained and rely on the compositions of the alloy, oxide and underlying coating [93]. According to this model, fluxing occurs due to the decomposition of oxides into corresponding cations and O^{2-} (known as acidic fluxing) or oxides with O^{2-} forming anions (referred to as basic fluxing).

In acidic fluxing, oxide ions are donated to the deposit melt through dissolving the oxide scale [93]:

$$MO = M^{2+} + O^{2-} \tag{2}$$

Acidic environments in molten deposits can be developed through two main processes, namely, alloy-induced and gas-phase acidic fluxing. Basic fluxing is achieved through the production of oxide ions in a Na_2SO_4 deposit. Such is obtained by removing S and oxygen from the residue through reactions with the alloy or underlying coating. Subsequently, the oxide scales (e.g., MO) produced can react with the oxide ions through reactions [93]:

$$MO + O^{2-} = MO_2^{2-} \tag{3}$$

A conventional model for LTHC was proposed by Luthra [111]. As suggested by the model, LTHC follows two stages, namely, formation of liquid-form sodium-cobalt sulphate and attack propagation through SO_3 migration through the liquid salt. In nickel-based alloys, the mechanism suggested by Shih and associates [112] for LTHC is sulphidisation.

4.4 Laboratory testing techniques

An alloy's resistance to hot corrosion can mainly be determined using four standard tests: the electrochemical, crucible, accelerated oxidation, and burner-rig [94, 113]. The crucible tests remain the most highly ranked test for hot corrosion, simply consisting of either suspending, depositing, or completely immersing the testing sample in molten salts at elevated temperatures, as presented in **Figure 3**.

As far as TiAl alloys are concerned, less work has been carried out to understand the hot corrosion behaviour of such alloys [114–116].

Gas turbine environments can be precisely simulated by employing burner-rig tests [117, 118], shown in **Figure 4**. The salt is in aerosol or fog form and fuel oil/air is introduced into the testing chamber to generate the test environment [119]. Simmons et al. [120] indicated that hot corrosion is an electrochemical process since hot corrosion consists of electrochemical reactions in which the molten salt acts as the conductive media or electrolyte.

4.5 Prevention methods

Some of the approaches used to prevent hot corrosion include maintaining both fuel purity and composition, properly selecting structural alloys, employing coatings, cleaning hot parts and air filtering [94].

4.5.1 Fuel purity and composition

Initiation and propagation of hot corrosion are greatly affected by impurities such as vanadium (V), S, and various alkali earth metals [121]. This can be controlled by adding magnesium (Mg), Cr, barium and calcium to the combustion fuel

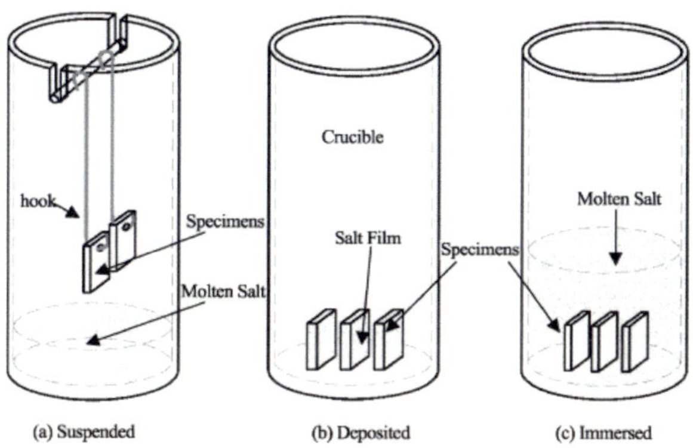

Figure 3.
Configurations used in hot corrosion crucible testing [114].

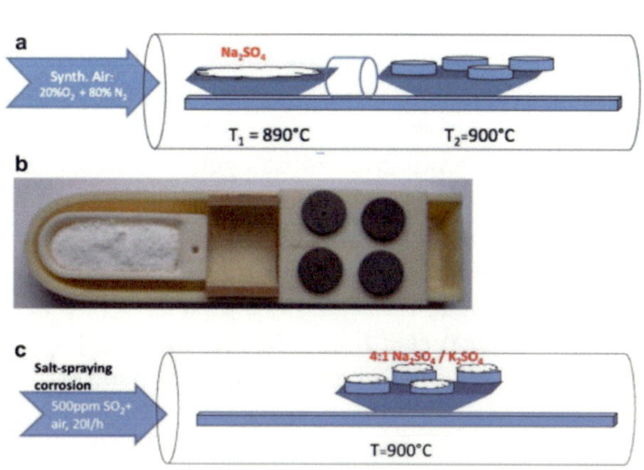

Figure 4.
Burner rig hot corrosion test schematic representation where (a) is an illustrates the experimental setup for Na_2SO_4(g) exposure, (b) is an image of the specimen plate for Na_2SO_4(g) tests in a crucible with the salt container and (c) is an ex-situ salt hot corrosion schematic diagram setup for experimental studies [119].

to decrease corrosion rate. The presence of zinc (Zn) in the form of anodes in the combustion fuel or as part of the protective coating can significantly reduce the occurrence of LTHC. According to Hancock and associates [122], Zn drastically reacts with chloride ions (i.e., when excess NaCl is available) and transfers the chloride to the gas-salt interface to transform to chloride gas via sulphidisation.

4.5.2 Proper alloy selection

The addition of Cr to superalloys has effectively reduced the occurrence of hot corrosion [123]. Historically [121, 124], Cr (15 wt.% for Ni-based and 25 wt.% for Co-based alloys) has been added to superalloys to reduce HTHC. Much related to TiAl alloys, Garip and Ozdemir [125] studied the effect of Cr, Mo and Mn on the cyclic hot corrosion behaviour, and subsequently reported the beneficial effects of Cr and Mn additions on the hot corrosion properties of the investigated samples. Cr's effect on corrosion resistance is attributed to the ability of Cr to form Cr_2O_3, stabilising the chemistry melt, preventing reprecipitation of the protective oxide scale. Contrarily, increased Cr additions to superalloys can compromise the high-temperature strength and ductility [113] by forming TCP phases. The alloy and oxide film adhesion has been reported to be improved by the addition of zirconium, yttrium, scandium, cerium and lanthanum [113]. Silicon (Si), platinum (Pt), hafnium, Ti, Al, and Nb [126] were also found to increase resistance to hot corrosion.

4.5.3 Protective coatings

Such as diffusion, overlay and thermal barrier (TBCs) coatings can be used on relatively resistant alloys to combat hot corrosion. An alloy's surface enrichment by Al, Si or Cr achieves diffusion coatings. Various aluminide diffusion coatings (i.e., PWA70, MDC3V, PWA62, TEW LDC2, Elbar Elcoat 360 and Chromalloy RT22) have been developed and can be alloyed with Pt to improve cyclic oxidation at high temperatures [127]. Overlay coatings, commonly referred to as M (base metal)–Cr–Al–Y coatings, are designed for LTHC and HTHC surface protection. Overlay coatings with low Cr-high Al coatings are used for HTHC protection, while high Cr-low Al coatings are used for LTHC [94]. TBCs protects the substrate from gaseous flow caused by heat and consist of an external ceramic usually zirconia) and an oxidation-resistant bond-coat overlay. Other coatings include intelligent coatings like RT22 (Pt-aluminide) and Sermetal 1515 (a triple-cycle Si-aluminide treatment), have been reported [127].

Inexpensive alternatives include oxide-based glass and glass–ceramic coatings [128, 129]. Oxide-based glass and glass–ceramic coatings exhibit a remarkable combination of properties such as excellent chemical inertness, high-temperature stability and superior mechanical properties, which effectively can mitigate deterioration caused by hot corrosion. The introduction of halogens on the surface of the alloy encourages the preferential formation of aluminium halides at elevated temperatures. The aluminium halides are then converted to thin, continuous, and protective alumina oxides. Fluorine provides the best oxidation protection [130]. Further examples of surface modifications coating and methods studied on γ-TiAl alloys include magnetron sputtering [131], laser cladding [132], sol–gel [133], pack cementation [134], chemical vapour deposition [135], slurry [136], ion implantation [137].

4.5.4 Cleaning hot parts and air filtering

Motoring washes can be flooded with plain water [121] to prevent hot corrosion using specified procedures in the maintenance manual for the specific engine model. Also, high-efficiency filters can be used to filter out air containing high sodium contents [138].

4.6 Hot corrosion studies for TiAl alloys

Although much work has been devoted to understanding the hot corrosion kinetics of Ni-based and Co-based superalloys, TiAl alloys emerged to have sparked much interest in recent years [1, 56, 57, 139]. Historically, reported works utilised alloys produced using conventional methods; however, more attention has recently shifted to AM routes [70, 73, 140–146]. Despite much devotion to improving structure–property relations of TiAls, little work has been reported on the hot corrosion of additively manufactured TiAl.

Garip and Ozdemir [147] produced an alloy to the nominal at.% composition of Ti-48Al-10Cr using electric current activated sintering and studied the hot corrosion kinetics of the alloys in Na_2SO_4 salt for 180 h at 700–900°C. A severe hot corrosion attack was observed at 900°C (refer to **Figure 5**), with a porous and loose layer consisting of $Na_2Ti_3O_7$, TiO_2, Al_2O_3 traces of TiS phase.

In a study led by Xiong et al. [67], bare alloys TiAl, TiAlNb, and Ti_3AlNb, were severely damaged after exposure at 750°C in $(Na, K)_2SO_4$ + NaCl melts as compared to those coated with enamel or TiAlCr. The corrosion mechanism was described to be much related to self-catalysis of sulphidisation and chlorination of metallic components. The initial mass loss observed is due to chloride volatility via metallic component chlorination. Of the alloys investigated, TiAlNb exhibited the best corrosion resistance due to adhesive Al_2O_3 enriched scale formation. Lastly, the degradation acceleration of sputtered TiAlCr coating was reported to be due to the chlorination of Cr and Al.

Additions of Nb and Si to traditional TiAl coatings were found to improve the hot corrosion resistance of a Ti-6Al-4 V alloy. In the stated work, Dai et al. [148] investigated the corrosion mechanisms on a mass loss basis following exposure at 800°C in a 75 wt.% Na_2SO_4 + NaCl salt mixture. Increasing single Nb additions deteriorated the hot corrosion resistance of the coating. Comparatively, increasing single Si additions continued to improve hot corrosion resistance. However, additions of both Nb and Si simultaneously showed better resistance to corrosion than single element additions. The corrosion protection of both Nb and Si (as seen in **Figure 6**) was related to SiO_2 and Al_2O_3 formation in the initial stages of hot corrosion. Secondly,

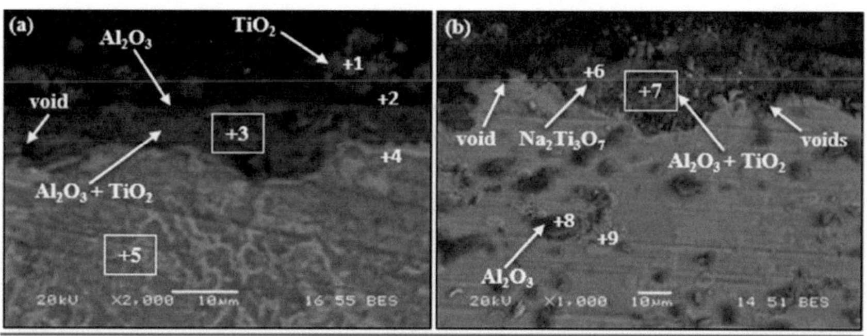

Element	1	2	3	4	5	6	7	8	9
O	67	49.5	60.5	24.9	3.8	66	63.3	60.9	30.5
Na	-	-	-	-	-	11.3	0.9	0.1	0.4
S	-	-	-	6.3	-	0.3	0.7	-	4.2
Ti	32.7	4.1	15.3	45.7	44.9	17.9	15.5	1.1	32.6
Al	0.2	45.3	14.6	16.8	42.8	4.3	19.1	37.9	28.9
Cr	0.1	1.1	9.6	6.3	8.5	0.2	0.5	-	3.4

Figure 5.
Cross-sectional SEM images showing oxide scale microstructures with EDS analysis points represented in at.%, after hot corrosion exposure at (a) 800°C and (b) 900°C for 180 h [147].

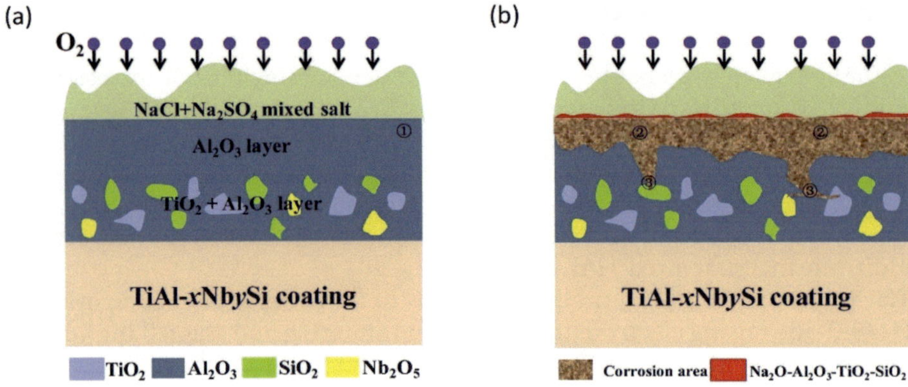

Figure 6.
Representative hot corrosion model of TiAl-xNbySi coating where (a) illustrates TiO_2 and Al_2O_3 formation and (b) shows an acidic dissolution of TiO_2 to form sodium titanates including $NaTiO_2$ and Na_2TiO_3 [148].

Si additions were reported to promote the formation of a Na_2O-Al_2O_3-TiO_2-SiO_2 enamel, hindering contact between the corrosive media and the oxide scales.

Tang et al. [149] studied the effect of enamel coatings on γ-TiAl against hot corrosion at 900°C. The enamel coating remained stable in the $(Na,K)_2SO_4$ melts, thus effectively protecting it against hot corrosion attack. Silicon-based coatings have also been shown to protect TiAl alloys. Rubacha et al. [150] evaluated the hot corrosion resistance of silicon-rich coated Ti-46Al-8Ta (at.%) alloy in NaCl, Na_2SO_4 and a mixture of the two salts. The formation of an amorphous SiO2 layer with TiO2 (rutile) and α cristobalite crystals enhanced the hot corrosion resistance of the TiAl alloy. Furthermore, Wu and colleagues [151] studied the hot corrosion resistance of a SiO_2 coated TiAl alloy in 75 wt.% Na_2SO_4 + 25 wt.% NaCl salt mixture at 700°C. The enhanced hot corrosion resistance of the TiAl alloy was attributed to the formation of a compact and adherent amorphous SiO_2 embedded with $Na_2Si_4O_9$ and cristobalite. The incorporation of Si in aluminide coatings has also provided long-term oxidation protection of γ-TiAl alloys at temperatures of 950°C by forming a continuous and uniform α-alumina oxide scale [152].

5. Oxidation

5.1 Definition

When metallic materials are exposed to elevated temperatures in air, oxidation occurs, resulting in the formation of oxide scales. The crystal structure of the individual metals significantly affects the oxidation rate of high-temperature applicative parts [153, 154].

5.2 Oxidation behaviour in TiAl alloys

The following reactions may occur when TiAl alloys are subject to an oxidising environment:

$$\tfrac{1}{2}Ti_{(s)} + O_{2(g)} = TiO_{(s)} \tag{4}$$

$$TiO_{(s)} + \tfrac{1}{2}O_{2(g)} = TiO_{2(s)} \tag{5}$$

$$2Al_{(s)} + \frac{3}{2}O_{2(g)} = Al_2O_{3(s)} \quad (6)$$

The ultimate oxidation resistance of alloyed TiAls is achieved by forming protective Al_2O_3, Cr_2O_3 and SiO_2 scales due to their outstanding thermal stabilities. In contrast, the unfavourable formation of porous TiO_2 with a high crack tendency is often observed [153]. Cobbinah et al. [155] found that 4 and 8 at.% Ta additions to Ti-46.5Al alloy promoted the significant formation of a consistent, non-porous Al_2O_3 layer at the metal-oxide boundary. Additionally, the layer operated as a diffusion barrier and preceded to outstanding oxidation resistance of the TiAl alloys.

In a study by Pan et al. [156], a comprehensive understanding is provided of the role of alloying on the oxidation resistance of TiAl alloys. Protection was related much to the formation of Ti_3Sn layer diminishing oxygen diffusion inwardly, promoted by Sn additions. Moreover, spallation resistance was enhanced by the Al_2O_3 oxide pegs providing a mechanical locking. The effect of cathodically electrodepositing a SiO_2 film on the oxidation resistance of a TiAl alloy was studied [157]. After 900°C exposure in air, the resultant alumina- and silicon-enriched glass-like oxide scale (in **Figure 7**) was reported, preventing oxygen diffusion leading to remarkably decreased alloy oxidation rates.

Surface modification of TiAl alloys via anodising has sparked interest in many high-temperature oxidation studies [158–161]. For instance, the oxidation behaviour and protection mechanisms of a TiAl alloy were studied by anodising in a methanol/NaF solution and produced an aluminium (Al)-and fluorine-enriched anodic film [162]. After 100 h exposure at 850°C, no evidence of cracking and spallation was displayed on the surface. The enhanced high-temperature oxidation resistance is mainly attributed to the halogen effect, generation of Al_2O_3 and oxidised Al–F species inhibiting external oxygen diffusion. Much effort has been devoted to developing coatings for γ-TiAl alloys, summarised in an evaluation by Pflumm et al. [130]. Amongst many available coating methods, Si-modified aluminide coatings produced via pack cementation have gained popularity. One such study [81] demonstrated that a continuous α-Al_2O_3 scale remained adherent after exposure to a temperature of 950°C for 3000 h.

5.3 Oxidation kinetics of TiAl alloys

When a metal operating at elevated temperatures is exposed to air, an oxide scale forms. As oxide scale formation proceeds, the metal's weight change can be plotted against time. Several laws such as linear, parabolic, logarithmic or cubic can be observed when studying oxidation kinetics [163]. In as far as TiAl alloys are

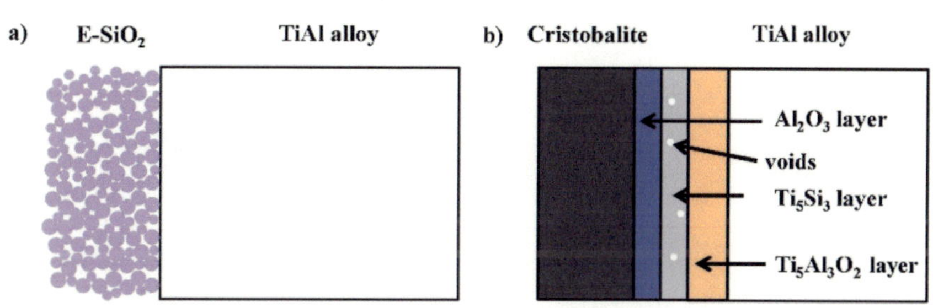

Figure 7.
Representation of a γ-TiAl alloy coated with E-SiO_2 film (a) and after thermal oxidation test (b) [157].

concerned, either linear or parabolic oxidation kinetics prevail. While the former offers no protection against high-temperature oxidation, the latter promotes diffusion-controlled oxide scale formation, improving much on the oxidation resistance of the base material. Parabolic oxidation follows and obeys the following law:

$$\left(\Delta m/A\right)^2 = k_p t \tag{7}$$

where Δm = change in weight (in mg), A = surface area (in cm^2), t = time (in sec) and k_p = parabolic oxidation rate constant (in $mg^2.cm^{-4}.sec^1$).

The optimum oxidation protection governed by the parabolic law often results in a thick and continuous TiO_2 and Al_2O_3 scale. As such, Swadźba et al. [48] investigated the short-term oxidation behaviour of a TiAl 48–2-2 alloy produced by AM at a temperature range of 750–900°C in air. At 900°C, a non-porous scale consisting of TiO_2, Al_2O_3 and nitrates, exhibiting parabolic oxidation (in **Figure 8**), was observed.

Garip [164] likewise studied the oxidation kinetics at 900°C in air for 200 h for TiAl alloys produced via pressureless and resistance sintering. Both alloys exhibited a nearly parabolic oxidation response, with oxidation rate constants of the pressureless sintered alloy of 0.6391 $mg^n\ cm^{-2n}\ h^{-1}$, 1.8 times higher than that of the alloy compacted using resistance sintering. Multi-layered oxide scales consisting of TiO_2 and Al_2O_3 were obtained.

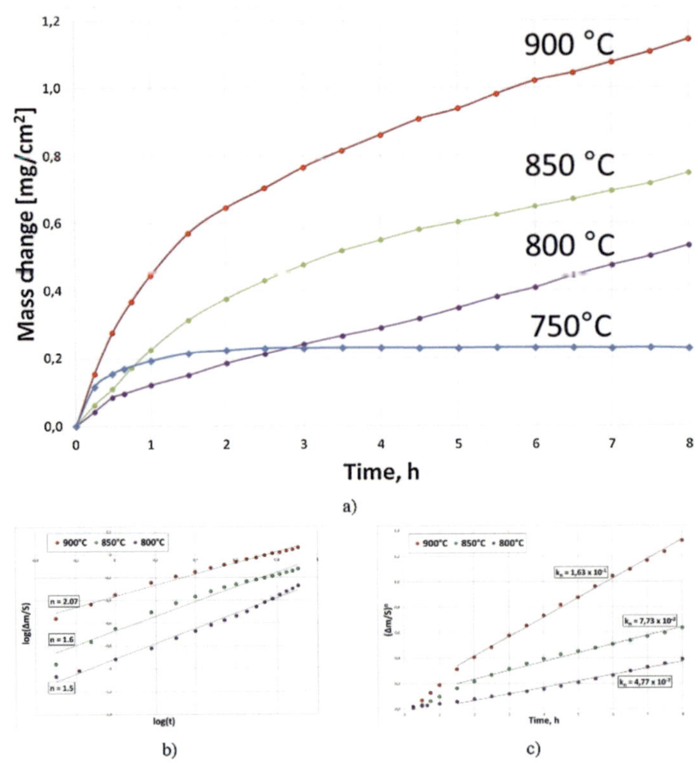

Figure 8.
Mass change against time plots for (a) oxidation rate constant of the AM produced TiAl 48–2-2 alloy and (b)–(c) the power-law constant – n extrapolation [48].

5.4 Effect of alloy modifications on the oxidation resistance of TiAl

Oxidation protection offered by forming a continuous Al_2O_3 scale followed by a multilayer of TiO_2 + Al_2O_3 is limited, unfortunately, to the maximum service temperature of ~830°C. Above this temperature, the protection potential presented by the oxide scales formed severely deteriorates, limiting the high-temperature application potential for structural components [165]. The current trend in research is to improve the oxidation resistance of TiAl through alloy modifications.

Nb is one element used in many research works [86–90] to improve the oxidation resistance of TiAl alloys. Al activities are promoted by Nb additions and accelerate protective Al_2O_3 oxide film formation, limiting oxygen diffusion into the alloy [166]. Also, the α_2 phase present in TiAl alloys is significantly decreased by Nb additions, decreasing its oxygen solubility [54]. Although Nb was primarily used for improving oxidation resistance [167], other high-temperature properties such as strength and creep resistance have been enhanced by the presence of Nb.

The creep resistance and the oxidation resistance of TiAl and its alloys can be enhanced by adding Si. The oxidation improvement is said to be achieved through the refinement of TiO_2 particles, inducing refined and compact TiO_2 scales on the surface [165]. Moreover, Si promotes Al diffusion into the oxide scale, stabilises Ti, reduces Ti^{4+} ions and impedes external Ti^{4+} ions diffusion [168].

The effect of adding molybdenum (Mo) alone to TiAls to improve on high-temperature oxidation is minimal. The protection of Mo-containing TiAl alloys is through the formation of inner oxide layers of TiO_2 and Ti_2AlMo near the substrate surface [165]. Unfortunately, Mo additions cannot alter the external oxide film formed (i.e., comprises of loose and porous TiO_2 scales) and its characteristics. It is recommended in practice that the improvement of high-temperature oxidation cannot be derived from adding Mo alone; instead, the combination of Mo with other alloys can have a beneficial effect on the alloys' resistance to oxidation [169].

Cr additions promote the formation of Cr_2O_3 oxides, which act as mass ion transport barriers [170], enhancing oxidation resistance. In addition, the Al content existing in the alloys can be significantly suppressed by Cr additions, promoting the formation of Al_2O_3 scales. Oxygen diffusion at elevated temperatures can be accelerated by Cr^{3+} ion doping in titanium oxide, improving oxygen vacancy concentration. Contrarily, the doping effect may impair the TiAl alloy's oxidation by making Ti^{4+} interstitially occupying TiO_2 sites, improving the potential energy with a noticeable decrease in diffusion activation energy, encouraging the diffusion of Ti^{4+} in TiO_2 [171].

Zirconium (Zr) additions can also enhance oxidation properties by altering the characteristics of the oxide formed during the primary stages of oxidation and promote oxide grain nucleation [172]. As a result, the refinement of the oxide particles occurs, which can hinder oxygen diffusion. Rare earth metals have been reported to enhance the oxidation resistance of TiAl alloys. As discussed in detail in a research paper by Dai et al. [165], the protection mechanism is contributed by grain refinement, substrate purification, oxide adherence improvement and promotion of Al selective oxidation.

6. Conclusions

The need for materials to give excellent mechanical properties under high temperatures and extreme conditions such as TiAl is in demand. The use of such alloys would mean a reduction in pollution and noise levels for aero-based engines due to improved thermal efficiencies. There are challenges in producing such alloys using the conventional arc and induction melt casting techniques due to the extremely high melting temperatures of the alloys. The AM route, particularly L-PBF, presents an opportunity to produce such alloys. What is critical in such trials is the operating parameters during processing. This has a direct influence on the performance and mechanical properties of the alloys so produced. Hot corrosion and oxidation of TiAl alloys are of great concern in gas turbine engines. Hot corrosion can be classified into HTHC and LTHC, with particular reference to mechanisms and characteristics. Protection control methods may result in fewer catastrophic failures. The hot corrosion process must be either totally prevented or detected early to avoid catastrophic failure. A sound understanding of oxidation mechanisms and kinetics of TiAl alloys makes it easier to tailor oxidation-resistant alloys by alloy modifications.

Acknowledgements

This research work is based on the research supported wholly/in part by the National Research Foundation of South Africa (Grant number 130004).

Conflict of interest

The authors declare no conflict of interest.

Author details

Ntebogeng Mogale*, Wallace Matizamhuka and Prince Cobbinah
Vaal University of Technology, Johannesburg, South Africa

*Address all correspondence to: ntebogeng.mogale@yahoo.co.za

References

[1] Mphahlele MR, Olevsky EA, Olubambi PA. Chapter 12—Spark plasma sintering of near net shape titanium aluminide: A review [Internet]. Cao G, Estournès C, Garay J, Orrù R, editors. Spark Plasma Sintering. Elsevier; 2019. p. 281-299. Available from: http://www.sciencedirect.com/science/article/pii/B978012817744000012X

[2] Reith M, Franke M, Schloffer M, Körner C. Processing 4th generation titanium aluminides via electron beam based additive manufacturing—Characterization of microstructure and mechanical properties. Materialia [Internet]. 2020;14:100902. Available from: https://www.sciencedirect.com/science/article/pii/S2589152920303185

[3] Bewlay BP, Nag S, Suzuki A, Weimer MJ. TiAl alloys in commercial aircraft engines. Mater High Temp. 2016;33(4-5):549-559.

[4] Clemens H, Smarsly W. Light-weight intermetallic titanium aluminides—Status of research and development. In: Advanced materials research. Trans Tech Publ; 2011. p. 551-556.

[5] Kim Y-K, Hong JK, Lee K-A. Enhancing the creep resistance of electron beam melted gamma Ti–48Al–2Cr–2Nb alloy by using two-step heat treatment. Intermetallics [Internet]. 2020;121:106771. Available from: https://www.sciencedirect.com/science/article/pii/S0966979519312014

[6] Cobbinah PV, Nzeukou RA, Onawale OT, Matizamhuka WR. Laser Powder Bed Fusion of Potential Superalloys: A Review. Metals (Basel) [Internet]. 2020 Dec 30 [cited 2021 Feb 5];11(1):58. Available from: https://www.mdpi.com/2075-4701/11/1/58

[7] Cui N, Wu Q, Yan Z, Zhou H, Wang X. The Microstructural Evolution, Tensile Properties, and Phase Hardness of a TiAl Alloy with a High Content of the β Phase. Materials (Basel) [Internet]. 2019 Aug 28 [cited 2021 Feb 5];12(17):2757. Available from: https://www.mdpi.com/1996-1944/12/17/2757

[8] Li J, Jeffs S, Whittaker M, Martin N. Boride formation behaviour and their effect on tensile ductility in cast TiAl-based alloys. Mater Des. 2020 Oct 1;195:109064.

[9] Liu H, Li Z, Gao F, Liu Y, Wang Q. High tensile ductility and strength in the Ti-42Al-6V-1Cr alloy. J Alloys Compd. 2017 Mar 25;698:898-905.

[10] Shu S-L, Tong C-Z, Qiu F, Zou Q, Jiang Q-C. Effects of ternary elements on the ductility of TiAl. Can Metall Q [Internet]. 2016 Apr 2 [cited 2021 Feb 5];55(2):156-60. Available from: http://www.tandfonline.com/doi/full/10.1080/00084433.2016.1169660

[11] Huang S-C, Hall EL. The effects of Cr additions to binary TiAl-base alloys. Metall Trans A. 1991;22(11): 2619-2627.

[12] Gu X, Cao F, Liu N, Zhang G, Yang D, Shen H, et al. Microstructural evolution and mechanical properties of a high yttrium containing TiAl based alloy densified by spark plasma sintering. J Alloys Compd. 2020 Apr 5;819:153264.

[13] Zhou H, Kong F, Wang X, Chen Y. High strength in high Nb containing TiAl alloy sheet with fine duplex microstructure produced by hot pack rolling. J Alloys Compd. 2017 Feb 25;695:3495-3502.

[14] Zhao ET, Niu HZ, Zhang SZ, Feng L, Yang SY. Microstructural control and mechanical properties of a β-solidified γ-TiAl alloy Ti-46Al-2Nb-1.5V-1Mo-Y. Mater Sci Eng A. 2017 Jul 31;701:1-6.

[15] Burtscher M, Klein T, Mayer S, Clemens H, Fischer FD. The creep behavior of a fully lamellar γ-TiAl based alloy. Intermetallics. 2019 Nov 1;114:106611.

[16] Ding R, Chiu Y, Chu M, Paddea S, Su G. A study of fracture behaviour of gamma lamella using the notched TiAl micro-cantilever. Philos Mag [Internet]. 2020 Apr 17;100(8):982-97. Available from: https://doi.org/10.1080/14786435.2020.1714088

[17] Wimler D, Lindemann J, Clemens H, Mayer S. Microstructural Evolution and Mechanical Properties of an Advanced γ-TiAl Based Alloy Processed by Spark Plasma Sintering. Materials (Basel). 2019;12(9):1523.

[18] Cui N, Wu Q, Bi K, Xu T, Kong F. Effect of heat treatment on microstructures and mechanical properties of a novel β-solidifying TiAl alloy. Materials (Basel). 2019;12(10): 1672.

[19] Chlupová A, Heczko M, Obrtlík K, Dlouhý A, Kruml T. Effect of heat-treatment on the microstructure and fatigue properties of lamellar γ-TiAl alloyed with Nb, Mo and/or C. Mater Sci Eng A. 2020 Jun 1;786:139427.

[20] Chen L, Zhu L, Guan Y, Zhang B, Li J. Tougher TiAl alloy via integration of hot isostatic pressing and heat treatment. Mater Sci Eng A. 2017 Mar 14;688:371-377.

[21] Stendal JA, Eisentraut M, Imran M, Sizova I, Bolz S, Weiß S, et al. Accelerated hot deformation and heat treatment of the TiAl alloy TNM-B1 for enhanced hot workability and controlled damage. J Mater Process Technol. 2021 May 1;291:116999.

[22] Mphahlele MR, Olevsky EA, Olubambi PA. Spark plasma sintering of near net shape titanium aluminide: A review. In: Spark Plasma Sintering. Elsevier; 2019. p. 281-299.

[23] Campbell I, Diegel O, Kowen J, Wohlers T. Wohlers report 2018: 3D printing and additive manufacturing state of the industry: annual worldwide progress report. Wohlers Associates; 2018.

[24] Yap CY, Chua CK, Dong ZL, Liu ZH, Zhang DQ, Loh LE, et al. Review of selective laser melting: Materials and applications. Appl Phys Rev [Internet]. 2015 Dec 1;2(4):41101. Available from: https://doi.org/10.1063/1.4935926

[25] Doubenskaia M, Domashenkov A, Smurov I, Petrovskiy P. Study of Selective Laser Melting of intermetallic TiAl powder using integral analysis. Int J Mach Tools Manuf. 2018 Jun 1;129:1-14.

[26] ISMAEEL A, WANG C shan. Effect of Nb additions on microstructure and properties of γ-TiAl based alloys fabricated by selective laser melting. Trans Nonferrous Met Soc China (English Ed). 2019 May 1;29(5):1007-16.

[27] Kenel C, Dasargyri G, Bauer T, Colella A, Spierings AB, Leinenbach C, et al. Selective laser melting of an oxide dispersion strengthened (ODS) γ TiAl alloy towards production of complex structures. Mater Des. 2017 Nov 15;134:81-90.

[28] Gussone J, Garces G, Haubrich J, Stark A, Hagedorn YC, Schell N, et al. Microstructure stability of γ-TiAl produced by selective laser melting. Scr Mater. 2017 Mar 15;130:110-113.

[29] Zhou YH, Lin SF, Hou YH, Wang DW, Zhou P, Han PL, et al. Layered surface structure of gas-atomized high Nb-containing TiAl powder and its impact on laser energy absorption for selective laser melting. Appl Surf Sci. 2018 May 31;441:210-217.

[30] Fu PX, Kang XH, Ma YC, Liu K, Li DZ, Li YY. Centrifugal casting of TiAl exhaust valves. Intermetallics. 2008 Feb 1;16(2):130-138.

[31] Aguilar J, Schievenbusch A, Kättlitz O. Investment casting technology for production of TiAl low pressure turbine blades—Process engineering and parameter analysis. Intermetallics. 2011 Jun 1;19(6):757-761.

[32] Liu K, Ma YC, Gao M, Rao GB, Li YY, Wei K, et al. Single step centrifugal casting TiAl automotive valves. In: Intermetallics. Elsevier; 2005. p. 925-928.

[33] Paul JDH, Lorenz U, Oehring M, Appel F. Up-scaling the size of TiAl components made via ingot metallurgy. Intermetallics. 2013 Jan 1;32:318-328.

[34] Zhang J, Jing Y, Fu M, Gao F. Microstructure optimization of ingot metallurgy TiAl. Intermetallics. 2012 Aug 1;27:21-25.

[35] Wang YH, Lin JP, Xu XJ, He YH, Wang YL, Chen GL. Effect of fabrication process on microstructure of high Nb containing TiAl alloy. J Alloys Compd. 2008 Jun 30;458(1-2):313-317.

[36] Trzaska Z, Couret A, Monchoux JP. Spark plasma sintering mechanisms at the necks between TiAl powder particles. Acta Mater. 2016 Oct 1;118:100-108.

[37] Kulkarni KN, Sun Y, Sachdev AK, Lavernia E. Field-activated sintering of blended elemental γ-TiAl powder compacts: Porosity analysis and growth kinetics of Al3Ti. Scr Mater. 2013 Jun 1;68(11):841-844.

[38] Wenbin F, Lianxi H, Wenxiong H, Erde W, Xiaoqing L. Microstructure and properties of a TiAl alloy prepared by mechanical milling and subsequent reactive sintering. Mater Sci Eng A. 2005 Aug 25;403(1-2):186-190.

[39] Gunasekaran J, Sevvel P, John Solomon I. Metallic materials fabrication by selective laser melting: A review. Mater Today Proc. 2020 Jun 3;

[40] Gao B, Peng H, Liang Y, Lin J, Chen B. Electron beam melted TiC/high Nb–TiAl nanocomposite: Microstructure and mechanical property. Mater Sci Eng A. 2021 Apr 15;811:141059.

[41] Todai M, Nakano T, Liu T, Yasuda HY, Hagihara K, Cho K, et al. Effect of building direction on the microstructure and tensile properties of Ti-48Al-2Cr-2Nb alloy additively manufactured by electron beam melting. Addit Manuf. 2017 Jan 1;13:61-70.

[42] Mohammad A, Alahmari A, Mohammed M, Renganayagalu R, Moiduddin K. Effect of Energy Input on Microstructure and Mechanical Properties of Titanium Aluminide Alloy Fabricated by the Additive Manufacturing Process of Electron Beam Melting. Materials (Basel) [Internet]. 2017 Feb 21 [cited 2021 Jul 13];10(2):211. Available from: http://www.mdpi.com/1996-1944/10/2/211

[43] Pradeep GVK, Duraiselvam Muthukannan, Katakam, Prasad S, Mohammad Ashfaq. Tribological Behavior of Additive Manufactured c-TiAl by Electron Beam Melting. Trans Indian Inst Met [Internet]. 1950;73. Available from: https://doi.org/10.1007/s12666-020-01950-8

[44] Xu Z, Ouyang W, Liu Y, Jiao J, Liu Y, Zhang W. Effects of laser polishing on surface morphology and mechanical properties of additive manufactured TiAl components. J Manuf Process. 2021 May 1;65:51-59.

[45] Kim YK, Hong JK, Lee KA. Enhancing the creep resistance of electron beam melted gamma Ti–48Al–2Cr–2Nb alloy by using two-step heat

treatment. Intermetallics. 2020 Jun 1;121:106771.

[46] Kim YK, Youn SJ, Kim SW, Hong J, Lee KA. High-temperature creep behavior of gamma Ti-48Al-2Cr-2Nb alloy additively manufactured by electron beam melting. Mater Sci Eng A. 2019 Aug 19;763:138138.

[47] Kastenhuber M, Klein T, Clemens H, Mayer S. Tailoring microstructure and chemical composition of advanced γ-TiAl based alloys for improved creep resistance. Intermetallics. 2018 Jun 1;97:27-33.

[48] Swadźba R, Marugi K, Pyclik. STEM investigations of γ-TiAl produced by additive manufacturing after isothermal oxidation. Corros Sci. 2020 Jun 1;169:108617.

[49] Bacos MP, Thomas M, Raviart JL, Morel A, Mercier S, Josso P. Influence of an oxidation protective coating upon hot corrosion and mechanical behaviour of Ti-48Al-2Cr-2Nb alloy. Intermetallics. 2011 Aug 1;19(8):1120-1129.

[50] Mitoraj-Królikowska M, Godlewska E. Hot corrosion behaviour of (γ + α2)-Ti-46Al-8Nb (at.%) and α-Ti-6Al-1Mn (at.%) alloys. Corros Sci. 2017 Feb 1;115:18-29.

[51] Godlewska E, Mitoraj M, Leszczynska K. Hot corrosion of Ti-46Al-8Ta (at.%) intermetallic alloy. Corros Sci. 2014 Jan 1;78:63-70.

[52] Qu SJ, Tang SQ, Feng AH, Feng C, Shen J, Chen DL. Microstructural evolution and high-temperature oxidation mechanisms of a titanium aluminide based alloy [Internet]. Vol. 148, Acta Materialia. 2018. p. 300-10. Available from: http://www.sciencedirect.com/science/article/pii/S1359645418301149

[53] Maliutina IN, Si-Mohand H, Sijobert J, Bertrand P, Lazurenko DV, Bataev IA. Structure and oxidation behavior of γ-TiAl coating produced by laser cladding on titanium alloy. Surf Coatings Technol. 2017 Jun 15;319:136-144.

[54] Kothari K, Radhakrishnan R, Wereley NM. Advances in gamma titanium aluminides and their manufacturing techniques. Prog Aerosp Sci. 2012;55:1-16.

[55] Huang SC, Chesnutt JC. Intermetallic Compounds. Principles and Practice, Vol. 2 Practice: Gamma TiAl and its Alloys. 1995;

[56] Mogale NF, Matizamhuka WR. Spark Plasma Sintering of Titanium Aluminides: A Progress Review on Processing, Structure-Property Relations, Alloy Development and Challenges. Metals (Basel) [Internet]. 2020 Aug 11 [cited 2021 Apr 20];10(8):1080. Available from: https://www.mdpi.com/2075-4701/10/8/1080

[57] Cobbinah P V, Matizamhuka WR. Solid-State Processing Route, Mechanical Behaviour, and Oxidation Resistance of TiAl Alloys. Adv Mater Sci Eng. 2019;2019.

[58] Ding J, Zhang M, Liang Y, Ren Y, Dong C, Lin J. Enhanced high-temperature tensile property by gradient twin structure of duplex high-Nb-containing TiAl alloy. Acta Mater. 2018 Dec 1;161:1-11.

[59] Appel F, Paul JDH, Oehring M. Gamma titanium aluminide alloys: science and technology. John Wiley & Sons; 2011.

[60] Yamaguchi M, Inui H, Ito K. High-temperature structural intermetallics [Internet]. Vol. 48, Acta Materialia. 2000. p. 307-22. Available from: http://www.sciencedirect.com/science/article/pii/S1359645499003018

[61] Clemens H, Mayer S. Design, Processing, Microstructure, Properties,

and Applications of Advanced Intermetallic TiAl Alloys. Adv Eng Mater [Internet]. 2013;15(4):191-215. Available from: https://doi.org/10.1002/adem.201200231

[62] Wimler D, Kardos S, Lindemann J, Clemens H, Mayer S. Aspects of Powder Characterization for Additive Manufacturing. Pract Metallogr [Internet]. 2018;55(9):620-36. Available from: https://doi.org/10.3139/147.110547

[63] Schimbäck D, Braun J, Leichtfried G, Clemens H, Mayer S. Laser powder bed fusion of an engineering intermetallic TiAl alloy. Mater Des. 2021 Mar 1;201:109506.

[64] Clemens H, Mayer S. Intermetallic titanium aluminides in aerospace applications—Processing, microstructure and properties. Mater High Temp [Internet]. 2016;33(4):560-570. Available from: http://search.ebscohost.com/login.aspx?direct=true&db=a9h&AN=118805116&site=eds-live

[65] Löber L, Biamino S, Ackelid U, Sabbadini S, Epicoco P, Fino P, et al. Comparison of selective laser and electron beam melted titanium aluminides. In: Conference paper of 22nd International symposium "Solid freeform fabrication proceedings", University of Texas, Austin. 2011. p. 547-556.

[66] Thomas M, Malot T, Aubry P, Colin C, Vilaro T, Bertrand P. The prospects for additive manufacturing of bulk TiAl alloy. Mater High Temp [Internet]. 2016 Jun 28;33(4-5):571-7. Available from: https://doi.org/10.1080/09603409.2016.1171510

[67] Zhang X, Mao B, Mushongera L, Kundin J, Liao Y. Laser powder bed fusion of titanium aluminides: An investigation on site-specific microstructure evolution mechanism. Mater Des. 2021 Mar 1;201:109501.

[68] Li W, Liu J, Zhou Y, Li S, Wen S, Wei Q, et al. Effect of laser scanning speed on a Ti-45Al-2Cr-5Nb alloy processed by selective laser melting: Microstructure, phase and mechanical properties. J Alloys Compd. 2016 Dec 15;688:626-636.

[69] Caprio L, Demir AG, Chiari G, Previtali B. Defect-free laser powder bed fusion of Ti–48Al–2Cr–2Nb with a high temperature inductive preheating system. J Phys Photonics [Internet]. 2020 Feb 12;2(2):024001. Available from: https://iopscience.iop.org/article/10.1088/2515-7647/ab7080

[70] Shi X, Wang H, Feng W, Zhang Y, Ma S, Wei J. The crack and pore formation mechanism of Ti–47Al–2Cr–2Nb alloy fabricated by selective laser melting. Int J Refract Met Hard Mater. 2020 Sep 1;91:105247.

[71] Gussone J, Hagedorn YC, Gherekhloo H, Kasperovich G, Merzouk T, Hausmann J. Microstructure of γ-titanium aluminide processed by selected laser melting at elevated temperatures. Intermetallics. 2015 Jul 21;66:133-140.

[72] Sola A, Nouri A. Microstructural porosity in additive manufacturing: The formation and detection of pores in metal parts fabricated by powder bed fusion. J Adv Manuf Process [Internet]. 2019 Jul 1;1(3):e10021. Available from: https://doi.org/10.1002/amp2.10021.

[73] Abdulrahman KO, Akinlabi ET, Mahamood RM, Pityana S, Tlotleng M. Laser metal deposition of titanium aluminide composites: A review. In: Materials Today: Proceedings. Elsevier Ltd; 2018. p. 19738-19746.

[74] Cobbinah PV, Nzeukou RA, Onawale OT, Matizamhuka WR. Laser Powder Bed Fusion of Potential Superalloys: A Review. Metals (Basel). 2021;11(1):58.

[75] Polozov I, Kantyukov A, Popovich V, Zhu JN, Popovich A. Microstructure and mechanical properties of tial-based alloy produced by selective laser melting. In: METAL 2020—29th International Conference on Metallurgy and Materials, Conference Proceedings [Internet]. TANGER Ltd.; 2020 [cited 2021 May 3]. p. 1037-41. Available from: https://doi.org/10.37904/metal.2020.3604

[76] Gao P, Wang Z. Formability improvement, cracking behavior and control of Y-modified Ti–43Al–4Nb–1Mo-0.1B alloys produced by selective laser melting. J Alloys Compd. 2021 Feb 15;854:157172.

[77] Ismaeel A, Wang C. Effect of Nb additions on microstructure and properties of γ-TiAl based alloys fabricated by selective laser melting. Trans Nonferrous Met Soc China. 2019;29(5):1007-1016.

[78] Mahajan S, Chhibber R. Hot corrosion studies of boiler steels exposed to different molten salt mixtures at 950°C. Eng Fail Anal. 2019 May 1;99:210-224.

[79] Prashar G, Vasudev H. Hot corrosion behavior of super alloys. 2020; Available from: https://doi.org/10.1016/j.matpr.2020.02.226

[80] Zhang W-J, Sharghi-Moshtaghin R. Revisit the Type II Corrosion Mechanism. Available from: https://doi.org/10.1007/s11661-018-4755-4

[81] Kumar S, Kumar M, Handa A. Combating hot corrosion of boiler tubes—A study. Vol. 94, Engineering Failure Analysis. Elsevier Ltd; 2018. p. 379-395.

[82] Liu YD, Sun J, Pei ZL, Li W, Liu JH, Gong J, et al. Oxidation and hot corrosion behavior of NiCrAlYSi+NiAl/cBN abrasive coating. Corros Sci. 2020 May 1;167:108486.

[83] He DG, Lin YC, Tang Y, Li L, Chen J, Chen MS, et al. Influences of solution cooling on microstructures, mechanical properties and hot corrosion resistance of a nickel-based superalloy. Mater Sci Eng A. 2019 Feb 11;746:372-383.

[84] Naghiyan Fesharaki M, Shoja-Razavi R, Mansouri HA, Jamali H. Evaluation of the hot corrosion behavior of Inconel 625 coatings on the Inconel 738 substrate by laser and TIG cladding techniques. Opt Laser Technol. 2019 Apr 1;111:744-753.

[85] Smialek JL, Nesbitt JA, Gabb TP, Garg A, Miller RA. Hot corrosion and low cycle fatigue of a Cr2AlC-coated superalloy. Mater Sci Eng A. 2018 Jan 10;711:119-129.

[86] HU Y tao, ZHENG L, YAN H jie, WU L kui, LIN X jun, CAO F he, et al. Improving hot corrosion resistance of aluminized TiAl alloy by anodization and pre-oxidation. Trans Nonferrous Met Soc China (English Ed. 2021 Jan 1;31(1):193-206.

[87] Garip Y, Ozdemir O. Corrosion behavior of the resistance sintered TiAl based intermetallics induced by two different molten salt mixture. Corros Sci. 2020 Sep 1;174:108819.

[88] Chang JX, Wang D, Zhang G, Lou LH, Zhang J. Interaction of Ta and Cr on Type-I hot corrosion resistance of single crystal Ni-base superalloys. Corros Sci. 2017 Mar 1;117:35-42.

[89] Cockings HL, Cockings BJ, Harrison W, Dowd M, Perkins KM, Whittaker MT, et al. The effect of near-surface plastic deformation on the hot corrosion and high temperature corrosion-fatigue response of a nickel-based superalloy. J Alloys Compd. 2020 Aug 15;832:154889.

[90] Nabavi B, Goodarzi M, Khan AK. Metallurgical effects of nitrogen on the microstructure and hot corrosion

behavior of Alloy 718 weldment. Mater Charact. 2019 Nov 1;157:109916.

[91] Aung NN, Liu X. Effect of SO2 in flue gas on coal ash hot corrosion of Inconel 740 alloy—A high temperature electrochemical sensor study. Corros Sci. 2013 Nov 1;76:390-402.

[92] Patel NS, Pavlík V, Boča M, Bo M. Critical Reviews in Solid State and Materials Sciences High-Temperature Corrosion Behavior of Superalloys in Molten Salts-A Review High-Temperature Corrosion Behavior of Superalloys in Molten Salts-A Review. 2016; Available from: https://www.tandfonline.com/action/journalInformation?journalCode=bsms20

[93] Nicholls JR, Simms NJ. Gas turbine oxidation and corrosion. In: Shreir's Corrosion. Elsevier; 2010. p. 518-540.

[94] Eliaz N, Shemesh G, Latanision RM. Hot corrosion in gas turbine components. Eng Fail Anal. 2002 Feb 1;9(1):31-43.

[95] Sulfidation: Turbine Blade Corrosion | Aviation Pros [Internet]. [cited 2021 Apr 14]. Available from: https://www.aviationpros.com/engines-components/aircraft-engines/turbine-engine-maintenance/article/10378159/sulfidation-turbine-blade-corrosion

[96] Part II: Corrosion Topics. In: Corrosion Atlas Case Studies. Elsevier; 2020. p. xliii–lxviii.

[97] Pettit F. Hot corrosion of metals and alloys [Internet]. Vol. 76, Oxidation of Metals. Springer; 2011 [cited 2021 Apr 15]. p. 1-21. Available from: https://link.springer.com/article/10.1007/s11085-011-9254-6

[98] LUTHRA KL. LOW TEMPERATURE HOT CORROSION OF COBALT-BASE ALLOYS—1. MORPHOLOGY OF THE REACTION PRODUCT. Met TRANS A [Internet]. 1982 [cited 2021 Apr 15];13(N 10):1843-52. Available from: https://link.springer.com/article/10.1007/BF02647841

[99] LUTHRA KL. LOW TEMPERATURE HOT CORROSION OF COBALT-BASE ALLOYS—2. REACTION MECHANISM. Met TRANS A. 1982;13(N 10):1853-1864.

[100] Lutz BS, Alvarado-Orozco J M, Garcia-Fresnillo L, Meier GH. Na2SO4-Deposit-Induced Corrosion of Mo-Containing Alloys. Present Address AUTOVISION Serv. 2017;88:599-620.

[101] Alvarado-Orozco JM, Garcia-Herrera J E, Gleeson B, Pettit F S, Meier G H, Meier GH. Reinterpretation of Type II Hot Corrosion of Co-Base Alloys Incorporating Synergistic Fluxing. Oxid Met [Internet]. 2018;90:527-53. Available from: https://doi.org/10.1007/s11085-018-9853-6

[102] Gheno T, Gleeson B. On the Hot Corrosion of Nickel at 700°C. Oxid Met. 84.

[103] Sumner J, Encinas-Oropesa A, Simms NJ, Nicholls JR. Type II Hot Corrosion: Kinetics Studies of CMSX-4.

[104] Haight H, Potter A, Sumner J, Gray S. New Technique to Map Hot Corrosion Damage: CMSX-4 Example.

[105] Lortrakul P, Trice RW, Trumble KP, Dayananda MA. Investigation of the mechanisms of Type-II hot corrosion of superalloy CMSX-4. Corros Sci. 2014 Mar 1;80:408-415.

[106] Lillerud KP, Kofstad P. Sulfate-Induced Hot Corrosion of Nickel. Vol. 21, Oxidation of Metals. 1984.

[107] Alizadeh M, Izadi A, Fathi A. Sensitivity Analysis on Turbine Blade Temperature Distribution Using Conjugate Heat Transfer Simulation. J Turbomach [Internet]. 2013 Sep 20;136(1). Available from: https://doi.org/10.1115/1.4024637

[108] Taamneh Y. Thermal analysis of gas turbine disk integrated with rotating heat pipes. Case Stud Therm Eng. 2017 Sep 1;10:335-342.

[109] Goebel JA, Pettit FS. Na2SO4-1nduced Accelerated Oxidation (Hot Corrosion) of Nickel.

[110] Goebel JA, Pettit FS. The Influence of Sulfides on the Oxidation Behavior of Nickel-Base Alloys.

[111] Luthra KL. Mechanism of low temperature hot corrosion. High Temp Corros. 1983;507-512.

[112] Shih S, Zhang Y, Li X. Sub-melting point hot corrosion of alloys and coatings. Mater Sci Eng A. 1989 Nov 15;120-121(PART 1):277-82.

[113] Otsuka N, Rapp RA. Hot Corrosion of Preoxidized Ni by a Thin Fused Na2 SO 4 Film at 900°C. J Electrochem Soc [Internet]. 1990 Jan 1;137(1):46-52. Available from: https://iopscience.iop.org/article/10.1149/1.2086436

[114] Zhang K, Li Z, Gao W. Hot corrosion behaviour of Ti-Al based intermetallics. Mater Lett. 2002 Dec 1;57(4):834-843.

[115] Tang Z, Wang F, Wu W. Effect of a sputtered TiAlCr coating on hot corrosion resistance of gamma-TiAl. Intermetallics. 1999 Nov 1;7(11):1271-1274.

[116] Tang Z, Wang F, Wu W. Effect of Al2O3 and enamel coatings on 900°C oxidation and hot corrosion behaviors of gamma-TiAl. Mater Sci Eng A. 2000 Jan 15;276(1-2):70-75.

[117] Nicholls JR, Saunders SRJ. Comparison of hot-salt corrosion behaviour of superalloys in high and low velocity burner rigs. High Temp Technol [Internet]. 1989 [cited 2021 Apr 19];7(4):193-201. Available from: https://www.tandfonline.com/doi/abs/10.1080/02619180.1989.11753437

[118] Lai GY, Barnes JJ, Barnes JE. A Burner Rig Investigation of the Hot Corrosion Behavior of Several Wrought Superalloys and Intermetallics [Internet]. 1991. Available from: https://doi.org/10.1115/91-GT-021

[119] Knutsson P, Lai H, Stiller K. A method for investigation of hot corrosion by gaseous Na2SO4. Corros Sci. 2013 Aug 1;73:230-236.

[120] SIMONS EL, BROWNING G V, LIEBHAFSKY HA. Sodium Sulfate in Gas Turbines★. Corrosion [Internet]. 1955 Dec 1;11(12):17-26. Available from: https://doi.org/10.5006/0010-9312-11.12.17

[121] Stringer J. High-temperature corrosion of superalloys. Mater Sci Technol [Internet]. 1987 Jul 1;3(7):482-93. Available from: https://doi.org/10.1080/02670836.1987.11782259

[122] Hancock PJ, Hancock HA, Caley WF, Hollingshead RS. A review of recent studies of the role of zinc as an inhibitor of hot corrosion from molten sulphates. Mater Sci Eng A. 1989 Nov 15;120-121(PART 1):313-8.

[123] Otsuka N, Rapp RA. Effects of Chromate and Vanadate Anions on the Hot Corrosion of Preoxidized Ni by a Thin Fused Na2 SO 4 Film at 900°C. J Electrochem Soc [Internet]. 1990 Jan 1;137(1):53-60. Available from: https://iopscience.iop.org/article/10.1149/1.2086437

[124] Gurrappa I. Hot Corrosion Behavior of CM 247 LC Alloy in Na2SO4 and NaCl Environments. Vol. 51, Oxidation of Metals. 1999.

[125] Garip Y, Ozdemir O. Corrosion behavior of the resistance sintered TiAl based intermetallics induced by two different molten salt mixture. Corros Sci. 2020 Sep 1;174:108819.

[126] Bornstein NS, DeCrescente MA, Roth HA. The relationship between

relative oxide ion content of Na 2 SO 4, the presence of liquid metal oxides and sulfidation attack. Metall Trans. 1973;4(8):1799-1810.

[127] Nicholls JR. Designing oxidation-resistant coatings. JoM. 2000;52(1): 28-35.

[128] Shen M, Zhu S, Wang F. Formation kinetics of multi-layered interfacial zone between γ-TiAl and glass–ceramic coatings via interfacial reactions at 1000°C. Corros Sci. 2015 Feb 1;91: 341-351.

[129] Liao Y, Zhang B, Chen M, Feng M, Wang J, Zhu S, et al. Self-healing metal-enamel composite coating and its protection for TiAl alloy against oxidation under thermal shock in NaCl solution. Corros Sci. 2020 May 1;167:108526.

[130] Pflumm R, Friedle S, Schütze M. Oxidation protection of γ-TiAl-based alloys—A review [Internet]. Vol. 56, Intermetallics. 2015. p. 1-14. Available from: http://www.sciencedirect.com/science/article/pii/S0966979514002325

[131] Laska N, Braun R, Knittel S. Oxidation behavior of protective Ti-Al-Cr based coatings applied on the γ-TiAl alloys Ti-48-2-2 and TNM-B1. Surf Coatings Technol. 2018 Sep 15;349:347-356.

[132] Liu XB, Wang HM. Modification of tribology and high-temperature behavior of Ti–48Al–2Cr–2Nb intermetallic alloy by laser cladding. Appl Surf Sci. 2006 Jun 15;252(16):5735-5744.

[133] Gao J, He Y, Gao W. Electro-codeposition of Al2O3–Y2O3 composite thin film coatings and their high-temperature oxidation resistance on γ-TiAl alloy. Thin Solid Films. 2012 Jan 1;520(6):2060-2065.

[134] Xiang ZD, Rose SR, Datta PK. Codeposition of Al and Si to form oxidation-resistant coatings on γ-TiAl by the pack cementation process. Mater Chem Phys. 2003 May 26;80(2): 482-489.

[135] Liang W, Ma XX, Zhao XG, Zhang F, Shi JY, Zhang J. Oxidation kinetics of the pack siliconized TiAl-based alloy and microstructure evolution of the coating. Intermetallics. 2007 Jan 1;15(1):1-8.

[136] Goral M, Swadzba L, Moskal G, Hetmanczyk M, Tetsui T. Si-modified aluminide coatings deposited on Ti46Al7Nb alloy by slurry method. Intermetallics. 2009 Nov 1;17(11): 965-967.

[137] Donchev A, Richter E, Schütze M, Yankov R. Improvement of the oxidation behaviour of TiAl-alloys by treatment with halogens. Intermetallics. 2006 Oct 1;14(10-11):1168-1174.

[138] Nakamori M, Kayano I, Tsukuda Y, Takahashi K, Torigoe T. Hot corrosion and its prevention in high temperature heavy oil firing gas turbines. In: Materials science forum. Trans Tech Publ; 1997. p. 633-640.

[139] Güther V, Allen M, Klose J, Clemens H. Metallurgical processing of titanium aluminides on industrial scale. Intermetallics. 2018 Dec 1;103: 12-22.

[140] Wimler D, Lindemann J, Reith M, Kirchner A, Allen M, Vargas WG, et al. Designing advanced intermetallic titanium aluminide alloys for additive manufacturing. Intermetallics. 2021 Apr 1;131:107109.

[141] Polozov I, Popovich V, Razumov N, Makhmutov T, Popovich *A. Gamma-*Titanium Intermetallic Alloy Produced by Selective Laser Melting Using Mechanically Alloyed and Plasma Spheroidized Powders. In: TMS 2020 149th Annual Meeting & Exhibition Supplemental Proceedings. Springer; 2020. p. 375-383.

[142] Polozov I, Sufiiarov V, Kantyukov A, Razumov N, Goncharov I, Makhmutov T, et al. Microstructure, densification, and mechanical properties of titanium intermetallic alloy manufactured by laser powder bed fusion additive manufacturing with high-temperature preheating using gas atomized and mechanically alloyed plasma spheroidized powders. Addit Manuf. 2020 Aug 1;34:101374.

[143] Polozov I, Sufiiarov V, Kantyukov A, Popovich A. Selective Laser Melting of Ti2AlNb-based intermetallic alloy using elemental powders: Effect of process parameters and post-treatment on microstructure, composition, and properties. Intermetallics. 2019 Sep 1;112:106554.

[144] Sizova I, Sviridov A, Bambach M, Eisentraut M, Hemes S, Hecht U, et al. A study on hot-working as alternative post-processing method for titanium aluminides built by laser powder bed fusion and electron beam melting. J Mater Process Technol. 2021 May 1;291:117024.

[145] Reith M, Franke M, Schloffer M, Körner C. Processing 4th generation titanium aluminides via electron beam based additive manufacturing—Characterization of microstructure and mechanical properties. Materialia. 2020 Dec 1;14:100902.

[146] Gasper AND, Catchpole-Smith S, Clare AT. In-situ synthesis of titanium aluminides by direct metal deposition. J Mater Process Technol. 2017 Jan 1;239:230-239.

[147] Garip Y, Ozdemir O. Comparative study of the oxidation and hot corrosion behaviors of TiAl-Cr intermetallic alloy produced by electric current activated sintering. J Alloys Compd. 2019 Apr 5;780:364-377.

[148] Dai J, Zhang H, Sun C, Li S, Chen C, Yang Y. The effect of Nb and Si on the hot corrosion behaviors of TiAl coatings on a Ti-6Al-4V alloy. Corros Sci. 2020 May 15;168:108578.

[149] Tang Z, Wang F, Wu W. Effect of Al2O3 and enamel coatings on 900°C oxidation and hot corrosion behaviors of gamma-TiAl. Mater Sci Eng A. 2000 Jan 15;276(1-2):70-75.

[150] Rubacha K, Godlewska E, Mars K. Behaviour of a silicon-rich coating on Ti-46Al-8Ta (at.%) in hot-corrosion environments. Corros Sci. 2017 Apr 1;118:158-167.

[151] Wu LK, Wu JJ, Wu WY, Yan HJ, Jiang MY, Cao FH. Hot corrosion behavior of electrodeposited SiO2 coating on TiAl alloy. Corros Sci. 2020 Sep 1;174:108827.

[152] Swadźba R, Swadźba L, Mendala B, Witala B, Tracz J, Marugi K, et al. Characterization of Si-aluminide coating and oxide scale microstructure formed on γ-TiAl alloy during long-term oxidation at 950°C. Intermetallics. 2017 Aug 1;87:81-89.

[153] Lim HP, Yun W, Liew H, Jet G, Melvin H, Jiang Z-T, et al. materials A Short Review on the Phase Structures, Oxidation Kinetics, and Mechanical Properties of Complex Ti-Al Alloys. 2021; Available from: https://doi.org/10.3390/ma14071677

[154] Samal S. High-Temperature Oxidation of Metals. In: High Temperature Corrosion [Internet]. InTech; 2016 [cited 2021 Apr 21]. Available from: http://dx.doi.org/10.5772/63000

[155] Cobbinah PV, Matizamhuka W, Machaka R, Shongwe MB, Yamabe-Mitarai Y. The effect of Ta additions on the oxidation resistance of SPS-produced TiAl alloys. Int J Adv Manuf Technol. 2020;106(7-8): 3203-3215.

[156] Pan Y, Lu X, Hayat MD, Yang F, Liu CC, Li Y, et al. Effect of Sn addition on the high-temperature oxidation behavior of high Nb-containing TiAl alloys. Corros Sci. 2020 Apr 15;166:108449.

[157] Wu LK, Wu WY, Song JL, Hou GY, Cao HZ, Tang YP, et al. Enhanced high temperature oxidation resistance for γ-TiAl alloy with electrodeposited SiO2 film. Corros Sci. 2018 Aug 1;140:388-401.

[158] Friedle S, Pflumm R, Seyeux A, Marcus P, Schü Tze M. ToF-SIMS Study on the Initial Stages of the Halogen Effect in the Oxidation of TiAl Alloys. Available from: https://doi.org/10.1007/s11085-017-9779-4

[159] Mo MH, Wu LK, Cao HZ, Lin JP, Zheng GQ. Halogen effect for improving high temperature oxidation resistance of Ti-50Al by anodization. Appl Surf Sci. 2017 Jun 15;407: 246-254.

[160] Mo M, Wu L, Cao H, Lin J, Zheng G. Improvement of the high temperature oxidation resistance of Ti–50Al at 1000°C by anodizing in ethylene glycol/BmimPF6 solution. Surf Coatings Technol. 2016 Jan 25;286: 215-222.

[161] Wu LK, Xia JJ, Jiang MY, Wang Q, Wu HX, Sun DB, et al. Oxidation behavior of Ti45Al8.5Nb alloy anodized in NH4F containing solution. Corros Sci. 2020 Apr 15;166:108447.

[162] Wu L-K, Xia J-J, Cao H-Z, Liu W-J, Guang-Ya Hou ·, Tang Y-P, et al. Improving the High-Temperature Oxidation Resistance of TiAl Alloy by Anodizing in Methanol/NaF Solution. 2018;90:617-31. Available from: https://doi.org/10.1007/s11085-018-9858-1

[163] Appel F, David J, Paul H, Oehring *M. Gamma* Titanium Aluminide Alloys.

[164] Garip Y. Investigation of isothermal oxidation performance of TiAl alloys sintered by different processing methods. Intermetallics. 2020 Dec 1;127:106985.

[165] Dai J, Zhu J, Chen C, Weng F. High temperature oxidation behavior and research status of modifications on improving high temperature oxidation resistance of titanium alloys and titanium aluminides: A review [Internet]. Vol. 685, Journal of Alloys and Compounds. 2016. p. 784-98. Available from: http://www.sciencedirect.com/science/article/pii/S0925838816319351

[166] Raji SA, Popoola API, Pityana SL, Popoola OM. Characteristic effects of alloying elements on β solidifying titanium aluminides: A review. Vol. 6, Heliyon. Elsevier Ltd; 2020. p. e04463.

[167] Kothari K, Radhakrishnan R, Wereley NM, Sudarshan TS. Microstructure and mechanical properties of consolidated gamma titanium aluminides. Powder Metall. 2007;50(1):21-27.

[168] JIANG H ren, WANG Z lei, MA W shuai, FENG X ran, DONG Z qiang, ZHANG L, et al. Effects of Nb and Si on high temperature oxidation of TiAl. Trans Nonferrous Met Soc China (English Ed. 2008 Jun 1;18(3):512-7.

[169] Pflumm R, Donchev A, Mayer S, Clemens H, Schütze M. High-temperature oxidation behavior of multi-phase Mo-containing γ-TiAl-based alloys. Intermetallics. 2014 Oct 1;53:45-55.

[170] Wei DB, Zhang PZ, Yao ZJ, Liang WP, Miao Q, Xu Z. Oxidation of double-glow plasma chromising coating on TC4 titanium alloys. Corros Sci. 2013 Jan 1;66:43-50.

[171] Zhou C, Yang Y, Gong S, Xu H. Mechanism of Cr effect for improvement of oxidation resistance of

Ti-Al-Cr alloys. Acta Aeronaut Asronautica Sin. 2001;22(1):73-77.

[172] Gaddam R, Sefer B, Pederson R, Antti ML. Oxidation and alpha-case formation in Ti-6Al-2Sn-4Zr-2Mo alloy. Mater Charact. 2015 Jan 1;99:166-174.